The So

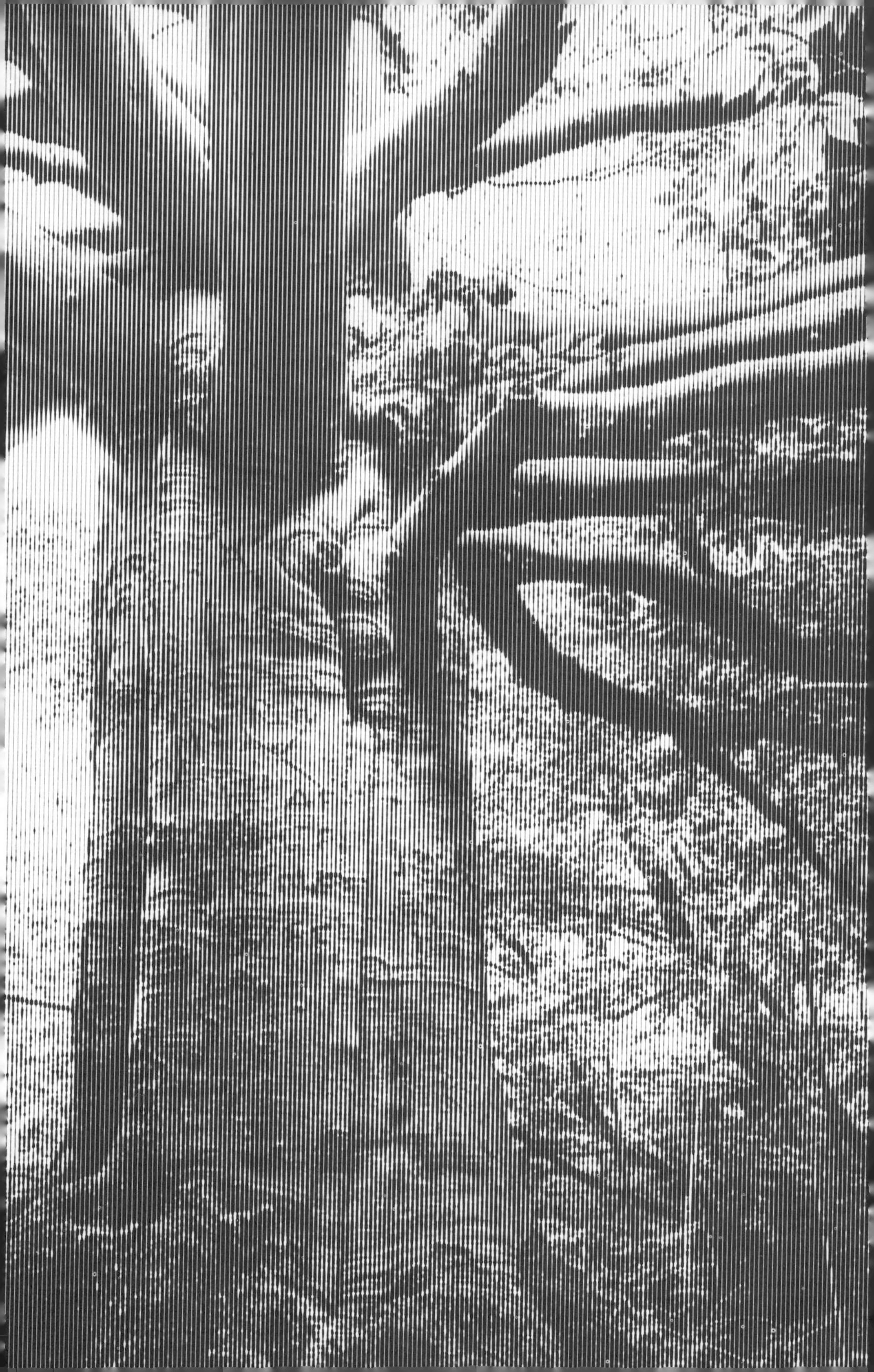

The Social Creation of Nature

Neil Evernden

The Johns Hopkins University Press
Baltimore and London

Published in cooperation with the Center for American Places, Harrisonburg, Virginia

Printed in the United States of America on acid-free paper
02 01 00 99 98 97 96 95 6 5 4 3

The Johns Hopkins University Press
2715 North Charles Street, Baltimore, Maryland 21218-4319
The Johns Hopkins Press Ltd., London

Photographs by Neil Evernden and Derek Evernden

LIBRARY OF CONGRESS CATALOGING-IN-PUBLICATION DATA

Evernden, Lorne Leslie Neil.
The social creation of nature / Neil Evernden.
p. cm.
"Published in cooperation with the Center for American Places, Harrisonburg, Virginia"—Verso t.p.
Includes bibliographical references (p. 167) and index.
ISBN 0-8018-4396-0 (acid-free paper). ISBN 0-8018-4548-3 (pbk.).
1. Natural history—Philosophy. 2. Philosophy of nature—History. I. Title.
QH331.E84 1992
333.7′01—dc20 92-6835

A catalog record for this book is available from the British Library.

For my parents, wife, and sons

Contents

Preface and Acknowledgments

At a time when issues of the environment seem to command universal attention, it is easy to believe that we are on the verge of confessing our collective sins and establishing a more benign relationship between ourselves and "nature." We are regaled daily with the promises of fresh solutions to the "environmental crisis"; yet it is difficult to suppress a sense of déjà vu.

It has been thirty years since Rachel Carson alerted us to the ecosystemic dangers of pesticide abuse, yet a rereading of *Silent Spring* leaves one with the feeling that little has changed but the names of the poisons. Even the much-lauded "sustainable development" sounds suspiciously like the system of environmental management that Gifford Pinchot advocated ninety years ago under the label "conservation." Indeed, many environmental tracts published early in the twentieth century could very nearly be used in contemporary classrooms.[1] One cannot avoid the sense that however much our environmental awareness has increased and our intentions to "save the earth" improved, at root nothing has changed. And if we genuinely care, we must surely ask why.

Those who raise such questions, however, regularly face charges of irrelevance or impracticality, for it is commonly expected that all worthwhile studies are to end with the optimistic provision of a practical solution: failure to do so ensures dismissal to the "ivory tower" category of literature. But this prejudice ignores two considerations. First, if, as many suspect, we continually propose the same solution under different names, there seems little to be gained by hearing it yet again. Indeed, there is good reason *not* to do so, given its apparent ineffectiveness and the suspicion that it may itself be the determinant of our futile efforts. In other words, perhaps the solution is determining the question: perhaps we inadvertently define our environmental *problem* as that to which the application of ever more technology is the only *solution*. And perhaps it is high time we reframed the problem and posed some different questions—or even asked if thinking in terms of problems and solutions is the only viable approach.

Second, the image of an ivory tower is inappropriate in current circumstances: no one is afforded the luxury of a safe tower above the environmental dilemma. Rather, such authors can more properly be regarded as occupying not a tower but a tunnel: an ivory tunnel, if you will, a viaduct to the roots of societal understanding. Yet even that may be too delicate an image, for what is carried in such tunnels is not rarefied vapors but the conceptual effluent of technological society that we flush away beneath our minds. Such sanitary niceties permit us to maintain the purity of our waking consciousness, but they also sustain us in our dangerous ignorance of our deeper motives and emotions.

Of course, the material that issues from such a tunnel is seldom composed of clear and distinct Cartesian particles whose identification and classification would provide a comforting illusion of certainty. Those who undertake the task of describing the fragments and aromas that drift past in the ivory gloom do so without the intention of posing a technical solution, but with the hope of sufficiently changing the landscape of assumptions that new questions can be asked. We are much the richer for the works of such writers as John Livingston, David Ehrenfeld, Carolyn Merchant, Morris Berman, Paul Shepard, and Max Oelschlaeger, to name only a few.[2] But although the ideas such authors present are highly relevant to the topic at hand, and would reward the attention of any reader, it is not my task to reconstruct their discussions here. Rather, I mean to pose an apparently simple question: what is this thing "nature" that we hasten to defend?[3] This may seem too obvious to bear asking, but it is often the things that seem most obvious that shield the greatest store of hidden assumptions. And indeed this is a topic so vast that I can aspire to provide only a thumbnail sketch of the development of the Western understanding of "nature."

To present a coherent overview of events spanning nearly five centuries, I have had to make some fairly sweeping generalities and to ignore a wealth of relevant detail. I hope, however, that by concentrating on the evolving *vision* of nature, rather than treating it strictly as an exercise in the history of natural philosophy or the history of science, we can better grasp the sense of nature that our ancestors bequeathed us, and better understand the implicit assumptions that today constrain our deliberations. Readers who wish to explore the topic in greater detail might wish to begin with the suggestions contained in the bibliographical essay at the end of this book.

Two other observations may be appropriate: first, the examination of "nature" must entail not simply the objects we assign to that category, but also the category itself: the *concept* of nature, its origins and implications. Second, in order to help discriminate between the "actual" and the "cultural," I have adopted the convention of speaking of "nature" when referring to the great amorphous mass of otherness that encloaks the planet, and to speak of "Nature" when referring specifically to the system or model of nature which arose in the West several centuries ago. Although this distinction may be initially confusing, I believe the need for it will become apparent.

What follows is not a work of philosophy, even though the contributions of philosophers certainly figure in it; neither is it biology, although it speaks of the natural world that is regarded as the domain of natural science; nor is it history, semiotics, or any of the multitude of disciplines that overlap the field of our concern. The generalist, lacking the security and guidance of disciplinary boundaries, must simply follow where the topic dictates. In this case the topic, nature, can take us virtually anywhere. Yet if we bear in mind that the task at hand is simply to think through the meaning we attach to this curious term, we may be able to follow the thread through the records of our distant ancestors without becoming excessively distracted by the enticing details that surround us. It is unlikely this could even be attempted, however, were we not blessed with some excellent guides. I have used many outstanding specialists as my "field assistants" in this attempt at an overview, and I hasten to add that this was done with respect and gratitude. I hope, however, that since nature, and indeed the environmental crisis, is too important to be left entirely to experts, I may be excused for presuming to bend their important insights to the task at hand. It seems unlikely that we can hope to "save nature" without first ascertaining just what it is we think we are attempting to save.

Perhaps I can make the story that follows more accessible by indicating in advance the pattern it is to take. In the first section (The Ambiguity of Nature), I outline the inherent volatility of the concept of nature and the ways in which that concept is used socially. Chapter 1 (The Social Use of Nature) illustrates some of the different versions of "nature" that are apparent in contemporary discourse, particularly in light of the environmental movement. Chapter 2 (Nature and Norm) traces the origins of the term itself and some of the concerns that arise from its ambiguity and appropriation. While we may collectively accept that nature is a distinct collection of entities about which we can achieve useful knowledge, nevertheless the *social* use of nature entails using slightly different versions of it in support of differing social ideals. Nature is, therefore, nowhere near as independent or as "given" as we like to suppose.

In the next section (The Creation of Nature), we deal with the development of the concept of nature over the course of several centuries. Naturally this is a highly condensed overview, and in order to facilitate a consideration of this enormously complex topic I have presumed to discern three distinct phases, which correspond roughly with the medieval period, the Italian Renaissance, and the empiricism of the seventeenth century. These are dealt with in chapters 3, 4, and 5, respectively (The Purification of Nature, From *nature* to *Nature,* and The Literal Landscape), and the overarching dualism that this development entails is treated separately in Chapter 6 (The Fragile Division).

This discussion of the development of the concept of nature prepares the way for the final section (The Liberation of Nature), in which I attempt to initiate a conversation about the natural world without the constraint of a conventional concept of what nature must be. That is, having attempted to demonstrate the highly cultural content of what is taken to be an objective entity, I suggest that we are justified in setting aside our current prejudices momentarily to see if, liberated from the conceptual shackles in which it is currently held, nature might be revealed in a different guise, and might, in that form, command a different attitude and response from us (Chapter 7, Nature and the Ultrahuman). This is not an attempt to formulate a new model of nature, but rather to indicate what the impediments to change are and how one might have to think about the human/nature relationship if alternatives are to be fostered.

Finally, I need to point out one stylistic feature intended to ease the course of exposition. I have occasionally employed what I think of as a "spiral" development, by which I mean that a theme is introduced in its most elemental form, and then re-examined at a later point after the context has been altered by the addition of relevant information. Hence, if any particular topic seems unclear at an early point in the text, the reader may be fairly confident that it will be clarified in due course. While this method does necessitate a certain amount of repetition, for which I must beg the reader's forbearance, I believe it is compensated for by a smoother conveyance over what could be treacherous terrain.

I must acknowledge the assistance of many people in this venture. I trust I have made my principal intellectual debts clear through quotation and footnoting, but the invisible assistance an author receives is never apparent in the text. I welcome the opportunity, therefore, to thank a sequence of colleagues, including John Livingston, William Fuller, Stephen Kline, William Leiss, and David Boag, for their stimulating conversation and friendship. I must particularly thank the following: George F. Thompson, president of the Center for American Places and a publishing consultant to the Johns Hopkins University Press, for initiating this project and patiently supporting its development; John Livingston, Bernard Lightman, and reviewers Yi-Fu Tuan and Max Oelschlaeger for their helpful suggestions; and Ann Waters for her careful and constructive editing. And not least, I must acknowledge the tolerance, support, and encouragement of my family over many years. Thanks, finally, to those "nonhuman others," whose existence is the sole justification for this endeavor.

Part 1

The Ambiguity of Nature

Custom is a second nature, which destroys the first one. But what is nature? Why is custom not natural? I greatly fear that nature may in itself be but a first custom, as custom is a second nature.

BLAISE PASCAL

1

The Social Use of Nature

To judge by the headlines and editorials of contemporary news media, there is a widespread sense that the whole of nature has become imperiled through profligate waste and human mismanagement. Although this state of affairs has not arrived unheralded, its emergence has been agonizingly slow. Certainly, it has been known at least since George Perkins Marsh's *Man and Nature* was published in 1864. Furthermore, it has been restated and embellished by a succession of concerned writers.[1] Yet some conceptual integument seems to have prevented its penetration into societal consciousness until very recently. Today, if we are to believe the proclamations of prophets and politicians, full realization has arrived. Nature is perceived to be in danger, and it is up to us to devise the means to its salvation.[2]

Given the current preoccupation with environmental affairs, it is often surprising to look back a few years and observe the apparent indifference of many to the impending crisis.

Even the ubiquitous term *pollution* did not acquire its current connotation and popularity until quite recently. The change may have been precipitated by Rachel Carson's heroic efforts to expose the folly of relying on biocides as a solution to even the most trivial of irritants. In any event, we seem to have no doubts about its meaning or significance today, and no shortage of individuals or industries at whom to point an accusing finger. Given the widespread anxiety about the consequences of all forms of pollution, it seems surprising that anyone would risk attracting such accusation. Yet even a cursory overview of the environmental battlefield reveals a surfeit of such offenses. Perhaps more importantly, however, it would also reveal significant disagreement about the *consequences* of such abuse. Or more accurately, there has been an ongoing debate between the accusers and the alleged perpetrators about what actually constitutes pollution.

In his survey of the opinions of different sectors of British society, sociologist Stephen Cotgrove detected some interesting differences in the apprehension of environmental risk. Two of his categories showed wide divergence: the "environmentalists" (composed of a sample drawn from membership lists of the Conservation Society and the Friends of the Earth), and the "industrialists" (selected from *Business Who's Who* and *Who's Who of British Engineers*). As one would expect, the environmentalists perceived considerably more environmental danger than did the industrialists. But what is interesting is that the latter group does not seem to be deliberately acting in an irresponsible way, but rather seems not to perceive significant risk at all.

If pollution is regarded as a matter of empirical fact, it may seem odd that such disagreements can persist. But since pollution involves questions not only of concentrations but also of consequences, even "hard" evidence is inevitably open to interpretation—hence the frequent spectacle of contradicting experts. Equally significant, however, is our tendency to treat pollution as a purely material phenomenon, a bias that tends to establish arbitrary boundaries to environmental debate.

We must bear in mind that the current understanding of pollution is just that: the current understanding. Yet there is no reason to limit the definition to physical abuse alone. The dictionary definition is much broader and entails "uncleanness or impurity caused by contamination (physical or moral)."[3] Our attention to physical pollution may distract us from the fact that much of the debate is over the perception of moral pollution. For example, while voicing their opinions about how many parts per billion of a toxin are "acceptable," both environmentalists and industrialists may be responding to a perceived instance of moral contamination. This emerges occasionally when one or the other makes predictions about future consequences, or about what "standard of living" ought to be protected. Environmentalists will assert that if the current action continues, our future well-being will be imperiled and our children will inherit a blighted planet. Cease, they say, and learn to live in a small-scale, cooperative society without the constant pressure for growth and transformation. Industrialists may reply that it is all very well for the impractical environmentalist to advocate such irresponsible action, but if their policies were ever to be put in place, our life-style would be in jeopardy, jobs would be lost, and food shortages would loom. To the environmentalists, what is at risk is the very possibility of leading a good life. To the industrialists, what is at risk is the very possibility of leading a good life. The debate, it appears, is actually about *what constitutes a good life.* The instance of physical pollution serves only as the means of persuasion, a staging ground for the underlying debate.

This is by no means a phenomenon unique to our times or our society. In *Purity and Danger,* the eminent anthropologist Mary Douglas makes it clear that the concept of pollution is a universal feature of human societies, and that our use of it is not fundamentally dissimilar to that of "primitives." All identify a contaminant, something that is out of place and hostile to the environment, as a danger to the well-being of individuals or society. On a personal basis, we respond to threats of pollution each time we wash our hands or clean the house: we attempt to get rid of dirt. But what is dirt? Not soil, but a contaminant, "matter out of place." Moreover, it is dangerous matter, at least when out of place. But significantly, something can only be out of place if there is a system of places to begin with.[4]

The system in question here is, of course, the environment. In order for there to be perceptible pollution, there must first be an understanding of systemic order, an environmental norm. Only then is it possible to detect something that is "out of place." But when we see differing ideals in confrontation, we observe different perceptions of pollution and risk. In a homogeneous society with a single environmental ideal, this misunderstanding is unlikely to occur. But in any society, we find ideas about pollution being used as a means of social control. "These danger-beliefs are as much threats which one man uses to coerce another as dangers which he himself fears to incur by his own lapses from righteousness. They are a strong language of mutual exhortation. At this level the laws of nature are dragged in to sanction the moral code."[5]

We witness just this use of pollution in the environmental debate. Ecology—the contemporary authority on nature's laws—is said to have revealed that such-and-such an action is dangerous to the environment, and that if it is continued both we and the environment will be imperiled. Hence, the polluter must be ordered to cease, lest he or she destroy us all. But notice that it is not just the environment that is at risk, but the very *idea* of environment, the social ideal of proper order. "The power which presents a danger for careless humans is very evidently a power inhering in the structure of ideas, a power by which the structure is expected to protect itself."[6]

The environmentalist and the industrialist possess differing ideas about the proper order of things. Moreover, since the debate is not simply about the physical contamination of nature but about the moral contamination of an ideal, there is also an underlying presupposition of "rightness" that adds a level of outrage and indignation to the debate.[7] The two groups support different ideals. Their debate, superficially about the contamination of drinking water, is, at root, really about what constitutes proper behavior. And to argue credibly, to enforce one's own view of pollution (and hence of environment and society), either side must be able to imply knowledge of what constitutes proper behavior. In a heterogeneous society, which tacitly embraces the notion of social relativity, we cannot articulate an absolute conception of proper behavior: there seems to be no external authority to which we can look for advice. But in this debate, there is ample opportunity to point to nature as the source of authority. That is, "unnatural" behavior can be discerned by the consequences to nature itself of environmental pollution, consequences that can be expressed in the manner most persuasive to contemporary society: quantitative measurement. Being able to determine the "parts per billion" of a contaminant enables the environmentalist to argue that pollution has indeed occurred, and thus to infer that the entire *position* of the polluter is untenable—the polluter has clearly done something "unnatural" and in so doing has placed nature, and ourselves, at risk. The polluter is condemned not only for a physical pollution but also, implicitly, for a moral pollution that is revealed by the physical pollution. Hence the highly charged emotional tone of much environmental debate: far more is at stake than the chemical composition of a river.

I must stress the significance of the underlying debate, because if there is to be a satisfactory response to our perception of environmental crisis, we must be aware of just what we think is happening—what is this "nature" that we believe to be at risk, and what exactly does that risk entail. Moreover, we must realize at the outset that the environmental crisis is as much a social phenomenon as a physical one.

As social beings, we inevitably engage in debates with our peers concerning the proper actions of societal members. Mary Douglas suggests that we most often resort to arguments based on time, money, nature, and God as our instruments of persuasion. All of these can be found in the environmental debate as well, but in this instance it is perhaps to nature that we turn most frequently for support. This may entail turning to ecology for an understanding of physical nature, but it can mean more than that; we often look to nature for insight into the state of our *social* environment as well. The latter practice, often concealed in the former, can be detected in the conclusions drawn and solutions proposed by the various advocates.

In public discussions of environmental affairs, ecology is frequently a rather loosely defined entity, often treated as the environmentalists' chief ally and occasionally even as a synonym for the natural environment. Indeed, the very plasticity of our concept of nature may be illustrated by the contrasting uses to which ecology is put. It is pressed into the service of a variety of social alternatives, ranging from various forms of

Aldo Leopold's land ethic to the bio-egalitarianism of Arne Naess and the eco-humanism of Murray Bookchin.* But exactly what is advocated is of less interest here than that ecology often functions as the exemplar of the natural and the healthy, and in so doing seems to indicate to us how we ought to re-orient our lives. Indeed, ecology will inevitably be so used if our understanding of ecology includes the establishment of norms as part of its function.

This new oracle, we seem to believe, will help us to feel our way into a healthier relationship with the world by revealing to us the "natural harmonies" that are essential to our survival and happiness. This popular understanding is not particularly attuned to the literature of ecology, in which one is more likely to find reference to "the myth of the balance-of-nature" and claims that "the success of the materialist revolution in other disciplines, particularly evolution and genetics, augurs well for ecology."[8] Even so, what ecology is may be less important than what it is believed to be, and the popular sense that nature knows best, and ecology knows nature, may be more significant than the musings of ecologists.

It is easy to understand how a more "convenient" ecology comes to be created in support of a program for social action. To a lesser extent, the same tendency can be found in the most careful of academic literature. For example, one very thoughtful author has suggested that "the basic concept behind an ecological ethic is that morally acceptable treatment of the environment is that which does not upset the integrity of the ecosystem as it is seen in a diversity of life forms existing in a dynamic and complex but stable interdependency."[9] He also refers to words that are used both conversationally and in a "strictly scientific, value-neutral" sense: balance, integrity, order, and health. He concludes that "stability and diversity

*I refer to three prominent figures in the literature of human-nature relationships: Aldo Leopold, a seminal figure in environmental thought and the originator of the well-known "land ethic," which was discussed in his collection of essays, *A Sand County Almanac* (New York: Oxford University Press, 1949); Arne Naess, the originator of the concept of "deep ecology" that has been discussed at length in Warwick Fox's recent overview of environmental thinking, *Toward a Transpersonal Ecology* (Boston: Shambhala, 1990); and Murray Bookchin, author of many books on "social ecology," such as *Toward an Ecological Society* (Montreal: Black Rose Books, 1980).

are both facts about a healthy ecosystem and values which we find in such a system."[10] Certainly, the terms are so used, and concepts such as stability and diversity have been regarded by some as ecosystemic virtues. But he neglects to mention other terms which share this joint societal/ecological usage, terms like competition, exclusion, exploitation, and survival. Such words are less likely to be seized upon by those aspiring to establish the groundwork for an ecological ethic or an environmentally benign life-style, yet they are every bit as much a part of ecology. Indeed, given the evolutionary emphasis in much ecological work, one could argue that competition forms a much more significant topic for research than does mutualism or balance. In choosing to ignore this somewhat dark side of ecology, even well-intentioned authors inadvertently create a new ecology to better serve their purposes—that is, to better substantiate their arguments.

If such authors can be criticized for embracing one subset of ecological concepts and ignoring the rest, many others could be accused of a reciprocal censorship, that is, of ignoring the "communal" terminology of ecology and emphasizing the exploitative. So far I have spoken of the use of ecology only by those in support of social reform. There is, however, a much heavier reliance on ecology by those who defend the status quo. I speak of the use of ecology in such officially sanctioned activities as environmental impact assessment, wildlife management, and land reclamation. While these may be useful in the immediate support of environmental integrity, they constitute a use of ecology in the service of technological and bureaucratic intervention. There is a tacit expectation that some form of environmental engineering must emerge that will facilitate continued growth with a minimum of environmental backlash. Ecology is to help us anticipate difficulties, so that alternative technologies can be forged to circumvent them. This role seems implicitly encouraged by ecologists who claim that "so long as we interest ourselves in prediction, we will make meaningful, practical statements about our environment. It is the sum of such statements which is the science of ecology, and this alone allows us some control over our collective future."[11] Indeed, one could cynically conclude that the role of the ecologist is to identify niches for humans to appropriate. Once the means of livelihood of a creature is known, it is up to us to mimic its adaptations through "appropriate technology" so that the human niche can be successfully expanded. To those who equate human biomass with the amount of good in the world, this might seem a noble enterprise (manifest destiny?). And all of us, dependent as we are on the holding back of succession so as to facilitate the monocultures we rely on, have a stake in the application of ecological knowledge.

These are two contrasting interpretations of the function of ecology. Undoubtedly they are caricatures of actual attitudes and assumptions, but they serve to illustrate the possibility of alternative uses of ecology, this contemporary nature-explainer that we expect to be "objective" and, of course, "value neutral." Persons with contrasting viewpoints can draw upon this discipline, one group regarding it as a revealer of the natural and proper, the other as a source of power and control (which it is natural for us to use). Each group believes its stance to be correct, and each expects endorsement from ecology. But before discussing the implications of this phenomenon, it might be instructive to consider a hypothetical use (or abuse) of ecology.[12]

The Alien Ecologist's Dilemma

Thus far, I have simply tried to remind the reader of the interaction between belief and discovery, and of the possibility of using a particular body of knowledge in support of alternative and possibly conflicting goals. Ecology cannot be presumed to be the exclusive ally of the environmental movement, for it provides information that can just as well be used to manipulate nature as to defend it. Nor is there any reason to assume that the findings of ecology provide credibility for environmentalists' social prescriptions. They can just as easily suggest a refined continuance of the status quo. In fact, ecology can even suggest a kind of behavior that neither environmentalist nor industrialist would be likely to regard as desirable. If, for the purpose of illustrating a point, I can ask the reader to let his or her fancy run free for a moment, the perils of looking to ecology for social instruction will become more apparent. I must caution, however, that although the example that follows contains various bits of ecological information to give it a semblance of authenticity, I do not mean to imply that it is anything more than a literary tool for dramatizing a point, or that any ecologist has actually suggested such an explanation.

Suppose that an uncommitted observer could study us as simply one organism among a multitude. To be able to make sense of this phenomenon, such a person would require some basic ecological understanding; and since we could scarcely expect an extraterrestrial to arrive with such knowledge, we will have to assume that our "alien" is an earth-born ecologist with selective amnesia. This disorder allows access to general principles of ecology, but to only fragmentary knowledge of individual species. We have, then, an educated observer of a strange biota.

In exploring this new-found world, our observer notices various organisms behaving in familiar ways, and is able to associate them with certain ecological roles. There are predators and parasites, opportunists and generalists, autotrophs and heterotrophs, creatures that are profligate and those that are self-regulating, and so on. Soon our observer discovers humans, without realizing any kinship with them. And with this discovery comes confusion, for this ambiguous creature seems to resist categorization. Its seemingly destructive behavior and extraordinary fecundity are surprising, and memory offers no parallels: the creature seems anomalous.

After a good sleep and a period of reflection, our ecologist begins to remember fragments of research, including some that indicated that unusually large numbers of a species may occur in certain situations, particularly in northern latitudes. Moreover, there were reports of some species appearing to dominate a community, only to shrink to insignificance a year or two later. And this memory in turn might encourage our ecologist to reflect on the puzzle of stability and instability, and on why some species seem able to self-regulate while others reveal their incompetence through fluctuation. Once this train of thought is initiated, our observer might also remember attempts made to find out what is "wrong" with species that are so poor at regulation as to condemn themselves to these boom-and-bust cycles of existence. And finally these musings might turn to the ingenious hypothesis that the species which seem most in jeopardy through this apparent regulatory flaw may in fact enjoy a more certain future than those which appear stable.[13]

In a partial remission of amnesia, our ecologist may recall the example of the Great Lakes fisheries, in which stable species were carefully managed for "maximum sustained yield." In spite of these apparently far-sighted managerial habits, the fish populations declined and never recovered, apparently because these "good" management practices had so consistently strained the propensities of the populations that a small additional stress was sufficient to alter their entire mode of being. That is, the species in question could not alter their "strategy" of existence, only the terms of their occupancy contract. Under pressure, they shifted from one "domain" to another, from one norm of population size to a lower one. Although these species were stable, they did not possess the resilience necessary to withstand a change in the conditions of existence. The only possible accommodation was to go on doing the same thing but on a more modest scale.

This kind of accommodation contrasts sharply with less stable species, which seem to respond to changing conditions on an ad-hoc basis. The way of being of such species, their strategy of existence, is uncertain but not fragile. In terms of population size, there is no "right way," no neat accounting of numbers. Such species are said to be unstable but resilient. In assessing the viability of the two types of accommodation, one might conclude that the important point is not so much how stable a species is within its domain but how vulnerable it is to a precipitate shift from one norm to another—that is, to a different configuration of existence. Since the Great Lakes fish were prone to such a shift, the management strategy adopted was precisely the wrong one. Maximum sustained yield, that most rational of our means of using "renewable resources," simply set these creatures up for a fall. If we believe that nature is manageable, it will appear to be so. But then again, it could turn its other, unmanageable face toward us when we least expect it.

All these are fleeting recollections to our detached observer, but they point inexorably toward consideration of this other category of creature, unstable but resilient. The ecologist thinks of the spruce budworm, scourge of the softwood forestry industry in Eastern Canada and cherished only by the occasional environmentalist (who is more concerned about the effects of spraying programs than about the creature itself). Actually, "outbreaks" of this insect had been quite rare until the assault by foresters commenced. There have been only six recorded eruptions since the early 1700s, apparently all the consequence of an unusual period of dry years coinciding with an abundance of mature balsam fir, the favored food of the budworm. Given those circumstances, a dramatic increase in accommodation and recruitment occurs for the budworm, and the population swells to a size that can affect the forest. The resulting decimation of the balsam fir then permits an increase in spruce and birch, thus refurbishing the spruce-fir-birch association upon which the budworm and a variety of other species depend. The persistence of this particular association appears to be linked to these periodic, selective rampages by the budworm. Of course, the forestry industry is not particularly interested in such niceties and would prefer a monoculture of economically significant species. Hence, the futile battle to prevent the budworm from completing its project of community rejuvenation. The unstable, resilient budworm is—through no "intent" of its own—the indispensable key to the existence of a whole community. The budworm is not a pest in this context, but a functioning part of the process we describe as an organic community.

At this point, our ecologist detects a straw to grasp at, a category that may accommodate the anomalous creature: perhaps this puzzling primate is, like the budworm, an agent of ecological renewal. Perhaps the dissemination and devastation that seem so characteristic are simply the manifestations of a kind of locust-being that is essential to the rejuvenation of aging systems. Perhaps this creature's strange fixation on the setting back of succession is indicative of its natural function. And perhaps what our observer has discovered is a "global budworm," a creature ideally suited to its role. Its essence is destruction, and in destroying it fulfills its destiny and its obligation to the biotic community. Unlike the minuscule and geographically contained efforts of its insect analogues, this exquisitely adapted organism executes its responsibilities on a planetary scale. Earth can be renewed through the devastation wrought by humankind. The mystery of the anomalous behavior is resolved, and our distraught ecologist can once again sleep contentedly.

The Lessons of a New Nature

Our observer might sleep well, but such conclusions could hardly have a soporific effect on those who take their environmental responsibilities seriously. If this behavioral taxonomy was accurate, then the one thing that is "natural" for human beings is to lay waste the countryside and not, as environmentalists might suggest, to enter into a long-term stable relationship that will arrest the rate of destruction. Given the role of global budworm, it would seem imperative that we set back succession and then decline to our former modest numbers. What would be inexcusable is accomplishing the decimation and then refusing to subside. Therefore, any accommodation that would permit continued existence at high population numbers would undermine all that has been accomplished, for the longer the retreat is delayed, the more permanent the environmental transformation would be. The budworm does not visit extinction on other species. At sustained high numbers, we inevitably do. And with nuclear weapons, we could succeed beyond our wildest dreams, perhaps even achieving a regression to a primary stage of succession—but of course the communities our activities might have revitalized would then no longer exist.

The persistence of huge populations is something that may be facilitated by the good works of the environmentalist, augmented by the application of biological technology. Indeed, perhaps the ultimate in biotechnological fixes would be the introduction of chloroplasts into the human epidermis (a prospect not so far-fetched as it might seem, given the advances in genetic engineering and the realization that chloroplasts, and perhaps other organelles as well, may have originated as independent organisms that colonized the cells of others), to permit direct utilization of sunlight.[14] This would free us from the tyranny of the green plant once and for all, thereby removing almost all limits to our global alterations. The use of biology seems so often to lead to the abuse of the biological. But if we were to be informed solely by the ravings of our detached ecologist, then we could be forced to conclude that the major enemies of earthly life are the environmentalists who seek a truce, while the devout developers with their apocalyptic vision of pavement and subdivisions represent the global budworm in its purest form. Given this discovery of purpose, an able philosopher might well manage to give voice to a "budworm ethic," in full confidence that the authority of ecology will attest to the naturalness of this venture. Clearly, environmental destruction is the proper and natural thing to do. A philosopher referred to earlier concluded that "the person who learns of the dynamic homeostasis of an undisturbed ecosystem, and, along with this ecological knowledge, feels an obligation not to harm that system, is aware of a moral obligation."[15] If that is so, surely a person who learns of the creative power of the budworm might also feel a moral obligation to destroy. Apparently it is a risky business, seeking advice from nature. The oracle changes with the questioner.

We have, then, three forms of belief about the action proper to human beings, all apparently justified by the insights of ecology. We can live "in harmony" with nature, which to some is clearly the "natural" thing to do; or we can expand our domain by direct competition with other species, which certainly seems (at least since Darwin) a "natural" enough thing to do; or we can endorse the overexploitation of nature in certain knowledge that through our destruction we are doing nature's work, just as we were "naturally" meant to do. Proponents of the first two possibilities would no doubt be united in rejecting the third, if in nothing else. But the point is not that the "budworm" scenario deserves serious consideration, but rather that nature justifies nothing, or anything. Ecology is today's official voice on natural matters, an institutional shaman that can be induced to pronounce natural whatever we wish to espouse. Ecology is, in this sense, simply being used as a blunt instrument to help implement particular life-styles or social goals. The "global budworm" story illustrates the possibility of justifying almost anything through reference to ecology and nature.

I doubt that any ecologist would be comfortable with the accounts I have given above, but as should be clear by now, science is not really the issue here. What matters is not what ecology is, but how it functions, how it is perceived and used—and perhaps *why* we seem compelled to assert such assumptions of "naturalness" at all. What ecologists do or think they do is, in this context, irrelevant. What is interesting is that ecology, and science in general, should be called upon in the defense of differing social ideals. This has been described in other settings, of course, perhaps most graphically in the debate surrounding the initiation of "sociobiology." The uses of biology that were revealed were quite striking, and not dissimilar to what we have been discussing. In that debate, it became apparent that natural selection was no longer taken to mean simply the transformation of species due to differential reproduction, but rather that natural selection was seen as the arena of competition. Animals were assumed to strive with one another to see who was more "fit," and hence who would "win" the right to increased reproduction. Selection is treated as a willed event, not the chance occurrence that Darwin described. Marshall Sahlins observed that "in the newer form of argument, selection indeed has lost its orienting power in favor of the maximization scheme of the individual biological subject. The structure of this argument transforms selection into the *means* by which DNA optimizes itself over the course of the generations. The orienting force of evolution is thus transferred from external life conditions to the organism itself." And since it comes to be accepted that natural selection entails the "expropriation of others' resources," then "Darwinism, at first appropriated to society as 'social Darwinism,' has returned to biology as a genetic capitalism."[16] Once again, nature is reformulated to become the kind of entity that will demonstrate the norms we wish to discover, and both nature and biology are pressed into the service of social reality.

The first step in establishing a new social order appears to be the construction of a new nature that will justify, even demand, its implementation. The trio of "ecologies" sketched above could support compliant behavior in an expanded community, dominant behavior as befits the pinnacle of competitive evolution, or destructive behavior suited to the embodiment of the environmental crisis. In each case, nature is tacitly asserted as the authority upon which appropriate behavior can be modelled. But how is it that nature, which we take to be an objective and neutral certainty, can be called into the service of differing ideals? Indeed, why should nature be so used at all? Is there some underlying ambiguity in our understanding of nature that leads to its intrusion into the realm of social ideals?

Such questions could probably be asked of any society. Yet perhaps no society other than our contemporary one has had so urgent a need to reconsider its motives in how it defines and uses nature. To untangle the knot of reasons and platitudes that binds us to our present understanding, we must consider not only the claims of the present-day interpreters but also those earlier transformations that have set the stage for the contemporary confrontation with the natural world. We must ask, in short, where "nature" came from.

2

Nature and Norm

Using *nature* to support contrasting ideals is certainly not an aberration unique to our own times. In fact, the ambiguous character of this term seems to encourage such use. This becomes apparent as soon as we examine the genesis of the word *nature.*[1] Its origins are veiled and its path of descent tortuous, but we can recognize many aspects of our own beliefs in the transformations of this word. C. S. Lewis notes that, on one hand, nature may be thought of "mainly as the real"—that is to say, as the nonfictional, in which case the laws of nature will be those that "enjoin what is really good and forbid what is really bad." Hence, a law of nature "is conceived as an absolute moral standard against which the laws of all nations must be judged and to which they ought to conform."[2] Given that, it is hardly surprising to find environmental advocates seeking instruction from the natural world, for this sense of nature as the ultimate basis of moral standards is admirably suited to the advocacy of a new and more generous treatment of the natural environment—and, for that matter, of other humans.

On the other hand, Lewis also describes a contrasting interpretation of this term as that which is the "least specifically human. The 'laws of *Nature*' on this view are inferred from the way in which non-human agents always behave, and human agents behave until they are trained not to." In other words, far from being the standard to which humans should aspire, these laws of nature constitute something which they are entreated to transcend. "The prime law of *nature,* thus conceived, is self-preservation and self-aggrandisement, pursued by whatever trickeries or cruelties may prove to be advisable."[3] It is Hobbes's "*natural law,*" and it is that abyss of brutality from which humans are said to have struggled to free themselves so as to permit existence in a civilized society. Variants of this interpretation are apparent in many contemporary exhortations to progress, which often describe the "ascent" of humanity to new planes of existence. In such a view, the domination of nature is not only a right but an obligation: nature is to be overcome, not preserved.

These contrasting understandings of the term *nature* therefore recommend two different and antithetical courses of action. And since each is by definition correct for its proponents, there can be no compromise. It is, as Cotgrove found, beyond the realm of reasoned debate. This kind of conflict, however, is merely the most obvious consequence of the varying uses of the word and concept *nature.* As we shall see, tacit understandings of "nature" figure prominently in many aspects of social discourse.

Lewis begins his discussion with the admission that the most common meaning of *natura* is simply what a thing is really like. To ask what is the nature of something is to ask about its character or essence. In this sense it has strong similarities with the term *kind,* which in much older usage was taken to mean what is typical or normal behavior for a person or thing. But by *nature,* and often nowadays also by *environment,* we also mean the world or a portion thereof, and in this usage it more closely resembles the Greek *phusis,* which itself has a complex etymology.

Phusis originally referred to what a thing is like, but for some reason—perhaps because it was used to question what the universe or "everything" was like—it eventually came to *mean* "everything." It is this expansive meaning as the conceptual container for the universe that was eventually transferred to *nature,* making it a term of fundamental importance. Lewis suggests that a small number of Greek thinkers effectively *invented* nature. Or rather, invented "Nature with a capital," or "*nature* in the dangerous sense," for, he claimed, of all the words he had analyzed, this is the one most likely to be employed where it is not required. Strictly speaking, there can be nothing that is not "nature"—it has no opposite. But "when *nature* . . . loses its purity, when it is used in a curtailed or 'demoted' sense, it becomes important."[4] In that demoted sense it is no longer "everything," and once the suggestion is made that there might be more to the world than "everything," more than their predecessors had known and had meant by *phusis,* an interesting transformation takes place. If *phusis,* or nature, is not all, then it may be thought of as just one thing, or one set of things. Furthermore, the pre-Socratics who invented nature "first had the idea (a much odder one than the veil of immemorial familiarity usually allows us to realise) that the great variety of phenomena which surrounds us could all be impounded under a name and talked about as a single object."[5] The possibility of having a *thing* called nature is as significant a development as a fish having a "thing" called water: where there was once an invisible, preconscious medium through which each moved, there is now an object to examine and describe.

As this possibility emerged, it was adapted in three different ways, which Lewis calls Platonic, Aristotelian, and Christian. The Platonic tradition presumed that what was other than or beyond nature was the eternal or archetypal forms, the true reality behind appearances. The Aristotelians regarded *phusis,* or nature, as the essential principle of change, the antithesis of the unchangeable things, gods and mathematics. And the Christian tradition essentially extended the Aristotelian concept so that God is regarded as the creator of nature, meaning that nature in this demoted sense is distinct from God, but also that He is related to it as a creator or artisan is to a piece of work, or as a master to a servant. It is principally this latter sense that has dominated, with modification, to the present, although aspects of the first two are also discernible in our attitudes to nature.

What is important to note, however, is that there *is* a metaphysic lying behind the simple existence of the word *nature.* It is not simply a description of a found object: it is also an assertion of a relationship. Furthermore, it plays a role, as we have already seen, in the daily life of a society through the social use we make of it. If there is nature, one can speak of things belonging to nature, or of being "natural." And if there are things that are natural, one can also speak of others as "unnatural." In the earlier use of nature as everything that is, this would not be possible. But if there can be something else, if nature is not everything, then some things may exist that are either beyond nature or *against* nature. In Lewis's treatment, "anything which has changed from its sort or kind *(nature)* may be described as *unnatural,* provided that the change is one the speaker deplores. Behaviour is *unnatural* or 'affected,' not simply when it is held to be a departure from that which a man's *nature* would lead to of itself, but when it is a departure for the worse."[6]

Notice that the change must always be one that is deplored, a negative departure from the norm—"unnatural" is a pejorative term. Furthermore, in being apart from or against nature, the "unnatural" will inevitably also mean "not natural," that is, not "given." Nature as given is that from which we start; change is inflicted from without. Hence, "for us, *nature* is all that is not man-made; the *natural* state of anything is its state when not modified by man. . . . We, as agents, as interferers, inevitably stand over against all the other things; they are all raw material to be exploited or difficulties to be overcome."[7] From the beginning, there is this sense of fundamental distinction between human and nature, along with a strange ambiguity that permits us to regard nature as the domain of both norms and forms.

However, just as "nature" has come to mean the world apart from human influence, it has also, thanks to its alternate meaning as the "way" or "essence" of things, been used to describe human "nature" or the "natural" element in humankind. To say that there is a natural element in humans may be a compliment, linking us with the greater sphere of nature.

Or, it may be a criticism, implying some lingering aspects of bestiality that must be overcome through civilization. This, of course, brings us back to our divergent "ecologies" and to the political usage of "natural." We may regard humanity as a part of nature and subject to absolute moral standards inherent therein, or as a creature able to overcome the remnants of nature within and build a better, supra-natural future for itself. And once again, the environmentalist and the industrialist divide, each armed with a semiotic club with which to impose a vision of an ideal existence. If nothing else, the budworm illustration provides a platform from which to observe both of these entrenched alternatives for what they are: entrenched alternatives—or, perhaps, as flip sides of a semantic coin that was minted in our distant past. Today, we can easily assume that we are making a rational decision to align ourselves with one side or the other in the environmental debate. But in a sense, we may simply be acting out roles created for us by playwrights or wordsmiths of old: our only apparent options are the ones they set forth, effectively blocking the way to further exploration. The semantic tools at our disposal may lend themselves to only certain types of conceptual excavation.

Nature as Sign

But we also "speak" of nature through images. It is no accident that nature figures prominently in the contemporary world of advertising, for that industry is largely concerned with effective communication, which inevitably means the effective use of signs.[8] Since advertising requires a powerful means of conveying favorable impressions of the product in question, the possibility of juxtaposing emotionally positive images with those of the product is irresistible. It is not surprising then that nature is used in advertising much as it is in the promotion of a new morality or world-view: as a visible manifestation of normalcy and health.

Nature has become a powerful part of our vocabulary of persuasion. But even that puts it too mildly, for it is often treated as the very realm of the absolute. To be associated with nature is to be placed beyond human caprice or preference, beyond choice or debate. When something is "natural" it is "the norm," "the way," "the given." This use of "nature" affords us a means of inferring how people ought to behave—including what objects they ought to associate with, that is, buy. Yet the authority of that usage stems in part from its confusion with the other major use, nature as the *material* given, nature as everything-but-us. In other words, the understanding of nature as the realm of external stuff, which is studied by science, lends an aura of objectivity and permanence to the understanding of nature as norm. The two mingle and interact so that we frequently lose sight of the distinction.

Indeed, one writer in the "communications" field, who has figured prominently in the emergence of the study now called semiotics, essentially equated nature with myth—not myth in the colloquial sense as superstitious or erroneous belief, or as primitive cosmology, but myth as an accepted story of the way the world is. Roland Barthes[9] treats myth as something of a second-order sign, and a sign, in turn, as the useful outcome of the juxtaposition of a "signifier" and a "signified." If we combine a signifier, a rose, for example, with a signified, such as passion, we have the rose as a meaningful sign, as a "passionified rose." Once established, that sign has, as it were, a life of its own: it "means." Similarly, on a second level we might take that sign as a new signifier which, in montage with a signified, creates a higher-order sign-system or "myth." And the danger of myth is that it will be taken not as a human creation but as an independent entity existing outside the realm of culture. It will be perceived, in other words, as nature, as a "factual system" when it is actually a "semiological system."[10] And when we are able to remove the impression of human agency from our description of the world and insinuate a natural reality, we will appear to be dealing with indisputable facts.[11]

But Barthes makes a surprising assertion in his discussion of mythology: he speaks of the need to "establish Nature itself as historical."[12] This seems contradictory, since we normally contrast the two as distinct and opposing realms. But he is speaking of the social use to which the concept of nature is put, and the mixing of the two realms is a phenomenon of considerable importance. Barthes is especially sensitive to the creation of a "nature" myth, since mythmaking seems to him to be the way in which social ideals—and social injustices—become entrenched. They are immune from analysis or criticism once they cease to appear as human concepts and instead become perceived as eternal givens. In other words, once something is perceived as lying in the realm of nature rather than in the realm of society or history, it seems beyond criticism. By definition, it has nothing to do with us: we are not its architects. Why criticize a sunrise or a frog? That's just the way the frog or the sunrise is, through nobody's fault. In fact, that is the way they were *meant* to be—an odd intuition, given that nature is no longer thought to have purposes or intentions. But this is the paradox: we resist the possibility of there being anything "human" in nature, including purpose and meaning, but then we proceed to use nature as a refugium for social ideals.

Once we can say, and believe, that a thing is natural, it is beyond reproach: it is now in the realm of the absolute. Through this process we are able to transform "the reality of the world into an image of the world, History into Nature. And this image has a remarkable feature: it is upside down." It appears that history or culture rests on nature, when in fact the reverse is true; nature becomes, in effect, a social creation, and "the passage from the real to the ideological is defined as that from an *anti-physis* to a *pseudo-physis,"* from a contrasting nature to a false or socially constructed nature.[13] This is not a conclusion unique to Barthes, although his elaboration of it is especially poignant. For example, in their groundbreaking study of the sociology of knowledge, Peter Berger and Thomas Luckmann suggest that "the real relationship between man and his world is reversed in consciousness. Man, the producer of a world, is apprehended as its product, and human activity as an epiphenomenon of non-human processes. Human meanings are no longer understood as world-producing, but as being, in their turn, products of the 'nature of things.' "[14]

Yet even if this is an irrepressible tendency of human consciousness, the particular instance that Barthes describes is especially significant. If, as he claims, we habitually make what we might call "anti-nature"—that is, culture or history, the antithesis of nature—into the entity we consider natural but which, since artificial, is really a false nature, a pseudo-nature, then we are no longer dealing with what we think we are. Nature as a physical reality does not enter into this usage at all, yet is seen to validate it. Nature, ambiguously regarded as a fact of direct experience and as the domain of the natural and inevitable, seems the ideal preservative for our cherished ideals.

Nature as Norm

This tendency to regard nature as a source of norms was also identified by Arthur Lovejoy and George Boas in their 1935 study, *Primitivism and Related Ideas in Antiquity.* Like Lewis, they identified ancient Greece as the source of the idea of nature and concluded that the availability of such a term made it almost inevitable that it be seized upon as the reference point for societal norms. Things that are seen to be variable, such as the contrasting mores of neighboring communities, are clearly creatures of custom, not given by nature. Therefore anyone seeking the truth, the "eternal" standards by which humans ought to live, would have to enquire which standards are given by nature. Hence the widespread interest in "primitives," who are often presumed to be living by those natural standards. But what is particularly interesting for our purposes is the inherent duality of the word *nature* and its remarkable facility for suggesting an independent source for social mores. When nature means the system as a whole, uninfluenced by humankind, it is an entity of extraordinary power—and great social utility. Attributing to any notion a connection with nature provides "an immediate certificate of legitimacy; its credentials need not be further scrutinized. Thus in any argument about ethical or political questions the word 'nature,' and consequently its ambiguity, often played a decisive part."[15] It still does, apparently: "To identify the objective ethical norm with those rules of conduct which are valid 'by' or 'according to' nature gave no logical answer to any concrete moral question; it was merely another way of saying that whatever is objectively right is objectively right, or that what is normal is normal."[16] This appears to be just what Barthes fears, that an empty argument—effectively simply an attribution of authorship—is sufficient to entrench social standards that may not be in the best interests of many members of the society.

But although this tendency can be discerned in many instances of social advocacy, it is perhaps most readily apparent in the case of environmentalists who claim to find in nature a new ethical standard of respect for the larger community; that is, an "environmental ethic" that could prevent wanton destruction of natural environments. However much one may sympathize with the intentions of such advocates, it is clear that they remain extremely vulnerable to this "demystification" of nature. That is, once it is pointed out that the norms they purport to see in nature are in fact of their own making, or even that the choices to be made are entirely subject to human judgment and not "natural" at all, their authority and their persuasiveness collapse.[17] Of course, the same may well be true of any other advocate, but as the group most explicit in its evocation of natural values, this one seems especially vulnerable to such criticism.

How else might the case for such social change be made? The environmental advocate sits on the horns of a dilemma: the time-honored technique of invoking the authority of nature has been essential to the presentation of a persuasive argument, and yet that technique is now vulnerable to charges of fraud. It seems an insoluble dilemma, perhaps the consequence of a linguistic trap. As Lovejoy and Boas observe, it was

> in a sense a historical accident that, when Greek thinkers had occasion to formulate the demand for objectively true principles, in contrast with mere conventions or subjective prejudices, in the realm of moral judgments, the term which they found made ready for them by the cosmologists and by the theorists about sense-perception was "nature." It is a historical accident in the sense that it is quite conceivable that the expression chiefly used for "the objective" might have been some other word—some much more inert and colorless and less ambiguous word—than "nature." If it had happened so, the history of European thought in many fields would doubtless have had a very different course.[18]

The mixing of meanings has made *nature* a potentially dangerous term, providing as it does endless opportunity for misunderstanding, deliberate or otherwise. Perhaps, then, if *nature* had not been so strangely adulterated, our understanding of it and of ourselves might now be quite different. And what we assume to be in nature makes all the difference. If nature is indeed red in tooth and claw, if it is "natural" to be competitive and "good" for both the species and society that the weak be eliminated, then the social structure that suggests itself—indeed, is demanded—is that which developed in the West. If, in contrast, it is "natural" to be compassionate and cooperative, then our recent history is a perversion, a distortion of our nature that is so hideous as to engender revolt, emotionally and politically. But of course the realization of this fundamental ambiguity, and of the pyramid of distortion that can be built upon the base "nature," forces us to question whether we may know "nature" at all. If our use of language has allowed us to conflate social norms and nature, then what might we be obscuring? Are we destined to always mistake our cultural stories for "nature?" What are we to do?

Scouring Nature

In commenting on the famous photographic exhibition of the 1950s, *The Family of Man,* created by Edward Steichen for the Museum of Modern Art, Barthes suggested that

> this myth of the "human condition" rests on a very old mystification, which always consists in placing Nature at the bottom of History. Any classic humanism postulates that in scratching the history of men a little, the relativity of their institutions or the superficial diversity of their skins . . . one very quickly reaches the solid rock of a universal human nature. Progressive humanism, on the contrary, must always remember to reverse the terms of this very old imposture, constantly to scour nature, its "laws" and its "limits" in order to discover History there, and at last to establish Nature itself as historical.[19]

"To establish Nature itself as historical" is a stunning project, certainly. If nature is simply a human-made entity, then it must not be permitted to be used as a safe-house for social injustices. Society must not be permitted to "naturalize" its failings, and to prevent that, the humanist must make it clear that *all* is history, all norms and ideals are of human creation and the product of human willing. Only thus can society be made safe; only thus can access be gained to all aspects of social belief. Nature must be established as historical, and as such fully vulnerable to human control.[20] Not just physical nature, the nature of resources, but also the "social" nature, nature as norm. It is another step in the conquest of nature: even the nature of thought must be domesticated. Yet if nature is to be seen as a subset of history, in what sense could it really be regarded *as* nature? And if we cannot regard it as an independent entity, with what can we replace it in our vocabulary of persuasion? From what realm are norms expected to arise? From reason alone? From the blueprints of "experts?" But who are they? To whom are we to entrust the engineering of the *concept* of nature?

Barthes is not, of course, denying the existence of physical nature, what we today call the "natural environment"; he is simply denying it normative content. But in his analysis he also reveals another assumption: that if we are to know nature *objectively*, free of social contamination, then we must employ a method of investigation so structured as to ensure this. We must examine a world free of subjective elements if we are to know the "real" nature. But although the concern with the "historicizing" of nature that Barthes exemplifies is largely a modern one, it is certainly not uniquely so. Nor is the suggestion that to obtain secure knowledge of nature one must first remove all human-generated content. Indeed, the exhortation to segregate all human qualities lies close to the roots of the modern conception of nature. It is to this requirement and its associated attitudes that we must now attend.

Nature as History

The revelation of our ability to construct "pseudo-natures" exposes two interesting things: one is our need for some external absolute to which we can look for authorization; the other is the requirement of a fundamental separation of "human" and "nature" for a modern world-view. Paradoxically, the two seem to go together: the more firmly we separate the two domains, and the more independent and self-willing we as humans seem to be, the more we are tempted to seek an absolute in the contrasting realm. If all societal norms are to be revealed as human creations, then all are vulnerable to revision or rejection: there is no certainty. Consequently, the temptation to secret a norm or two in the realm beyond human willing is considerable.

In discussing the use of nature in social communication, Barthes has tacitly raised what is perhaps the thorniest issue with which members of this society must contend, the admission of historical relativity. We all face, with alternating joy and repulsion, the idea that there is no absolute standard for social mores: "different strokes for different folks" has become the platitude of the day, and one is expected to quietly tolerate any and all behavior as long as one is not personally damaged by it. Of course, our acceptance of this belief is flawed, and we frequently criticize the behavior of others as "wrong" or "unnatural," or perhaps just "maladaptive," when we find it divergent from our own standards. Yet our anger is impotent: if all is relative, we really have no means by which to criticize and correct others, or to entrench our own "values"—the contemporary term for human-generated norms which we possess in place of absolute norms or "instincts." We live in chronic frustration, paying lip service to freedom and to the right of the individual to do as he or she chooses while at the same time lamenting libertine life-styles, the decline of public mores, crime in the streets, and generally all the fruits of societal freedom which do not harmonize with our own judgments of propriety. Perhaps even more challenging, though less commonly addressed, is the concomitant lack of purpose that we all experience. That is, the absence of external authority that makes possible this relativistic freedom also removes any given end for the project of human existence. This we tend to conceal by a conspiracy of silence, or by a somewhat hypocritical splicing of certain fragments of religious life onto the unrestricted freedom we expect in daily secular activities. But despite our attempts at camouflage, the crisis remains. As one of Walker Percy's characters observes, "This is not the Age of Enlightenment, but the Age of Not Knowing What to Do."[21]

In light of this, the tendency to practice the subterfuge of mythmaking is understandable. In practical terms, it may very well afford us some measure of comfort by legitimating a belief in the certainty of at least a few features of existence and a few behavioral norms. But in the long run it solves nothing, and has the added effect of drastically transforming—one might say contaminating—nature. That is, what we know as nature is what we have *constituted* as nature. Karl Löwith observed that

> so-called historicism would be harmless if it had merely historicized and relativized the so-called spiritual world. It made nature relative to us, with the effect that actually nothing natural was left over. In our scientifically organized world, naturalness is no longer the standard of nature. What still remains of natural things seems to be a mere leftover of that which has not yet been thoroughly subjected by man. This historical appropriation of the natural world is at the same time an estrangement from it.[22]

What does it mean to say that nothing natural has been left over in nature? That not only has God become relative to our judgments, but also nature? That with the Barthesian thought-police at work weeding nature from our cultural gardens, all but a few fragments have been domesticated? That whatever exists does so at the whim of humankind? Meaning is not, *may* not be, a part of nature. Meaning and purpose are solely the consequence of human thought: "the quest for meaning has become focussed in history." In fact, the transformation in thought has been so complete that this statement is not in the least surprising; where else could meaning come from but the human will? Yet our ancestors of only a few centuries ago could not have contemplated *that* prospect any more than we can imagine meaning deriving from some source outside ourselves. Löwith asks: "Why do we not ask for the meaning of all that exists, not alone through us but without our devices, by nature? Why does the natural light of the stars mean less to us—almost nothing—than a traffic light? Obviously because the meaning of a traffic light is in its purpose, while the light of sun, moon, and stars has no human and artificial purpose,"[23] and therefore no purpose at all. As we are no longer *in* nature conceptually, purpose is no longer intermingled with the stars and the birds. With the great divorce between humanity and nature, purpose and all other species of willing came to reside with us: we have sole custody of these offspring, and are pleased to regard them as the consequence of a virgin birth—perhaps the last miracle we have been able to accept.

Löwith concludes that with a change in our understanding of the world goes a change in our understanding of ourselves, of "the human condition," a phrase which he notes was to replace "the nature of man" in our discussion of our own affairs. The emptiness we feel and our apparent want of direction are inevitable correlates of our discovery of an infinite universe governed by chance. The *overcoming* of chance thus becomes our central obsession. It is interesting that the very term that Barthes criticizes—"the human condition"—is identified by Löwith with the establishment of a historicized nature. Recall Barthes's claim that the "myth of the 'human condition'" rests on a "mystification" that places nature at the bottom of history, thus encouraging the belief that "in scratching the history of men a little, the relativity of their institutions or the superficial diversity of their skins . . . one very quickly reaches the solid rock of a universal human nature." He rejects this, the position of "classic humanism," in favor of another: "Progressive humanism, on the contrary, must always remember to reverse the terms of this very old imposture, constantly to scour nature, its 'laws' and its 'limits' in order to discover History there, and at last to establish Nature itself as historical."[24]

Humanism

But what is this "progressive humanism" to which Barthes alludes? What is it that justifies the imperative of "scouring nature" to remove historical impurities? Humanism flourished initially as a Renaissance movement centered on classic texts, but it subsequently became concerned with humanity in general. In recent times it has taken on a somewhat different meaning, and is sometimes conflated with such terms as "humanitarian" or "humanities" or even "humaneness." Humanism, however, is more properly understood as a form of philosophy which places humanity at the center, displacing God, nature, and all other deities. By dictionary definition, it is "devotion to human interests; system concerned with human (not divine) interests, or with the human race (not the individual); Religion of Humanity."[25] Others elaborate slightly: "Humanism is also any philosophy which recognizes the value or dignity of man and makes him the measure of all things or somehow takes human nature, its limits, or its interests as its theme."[26] Most members of contemporary society would find themselves in sympathy with at least some of these themes, and indeed may take them entirely for granted. Unfortunately, the near-obsession with humanity alone makes the genuine valuing of the nonhuman next to impossible. "Human progress" becomes the sole legitimating principle.

Charles Hartshorne pointed out more than fifty years ago that in the best sense humanism is "simply an expression of an interest in man," but that in the worst sense "it is this interest become a monomania, excluding interest in anything else." If the only subject of worth is the human, there is very little incentive to attend to any other. Moreover, humanists generally "hold that, so far as we know, man is the highest type of individual in existence, and that therefore if there is any proper object of religious devotion, any real 'God,' it can only be humanity considered in its noblest aspirations and capacities, together with nature so far as expressed in and serviceable to humanity."[27]

That statement, written in 1937, could well have been designed as a preface to the modern environmental movement.[28] It demonstrates both the intensity of separation required between the human species and the rest of the world, and the virtual impossibility of considering questions of worth and value beyond the protected domain of humanity. Given the state of affairs that it describes, the difficulty that our society has in deciding on its position vis-à-vis "nature" is entirely understandable, as is the dismissal of the more "extreme" environmental arguments. An illustration of the difficulty which even the most thoughtful commentators have in dealing with the issue of human/nature interaction might be useful at this stage.

A respected Canadian journalist, Robert Fulford, recently wrote a column concerning the so-called animal rights[29] movement, and concluded that

> animal liberation challenges the idea that humans are superior, in any significant way, to animals. This breathtakingly audacious notion is never quite stated but is often

> implied in the writings of [Peter] Singer and others. In essence, it seems to me crankily misanthropic. One must have learned to hate one's own species a great deal in order to put it on the same level with creatures who cannot do what human beings, uniquely, can do—exercise their imaginations, create grandly, see themselves in history, come to some understanding of the world. Animal liberation, as now conceived, could have been born only among humans profoundly disappointed by their own species. In that sense, perhaps, it can be seen as a peculiarly extreme expression of the spirit of our time.[30]

With the exception of his suggestion that the mere expression of such ideas is indicative of the spirit of our times, his assessment seems to me to be entirely accurate. It can well be seen as "crankily misanthropic," in a particular sense. But it is curious that the advocacy of the well-being of any nonhuman creatures is taken to indicate a hatred of humanity. Why should concern for the whooping crane or the wolf be "read" as advocating the demise of humanity? Unless one envisages love as a commodity, which each individual possesses in limited supply to dispense to a set number of chosen targets, there is no apparent risk to humanity in a proclamation of affection for wildlife. Why, then, is the charge of misanthropy so readily leveled? The answer may have little to do with logic, or even with the espousal of the cause of the whooping crane. Rather, any expression of concern for the bird is taken to be evidence of a deep and dangerous heresy: the failure to pay adequate homage to the One True Deity, Humanity. In expressing concern for nonhumans, the individual reveals a failure in devotion—not toward other people, for there is no evidence that nature advocates are in any way deficient in their treatment of their fellow humans,[31] but toward the abstraction "humanity." For it is this idea, the notion of a collective entity of cosmic importance, that is the base of the current faith, a faith that arises simultaneously with the extraction of all significance from the world and, ultimately, even from God, to be vested solely in the new object of worship. When that abstraction becomes accepted as the sole source of everything of worth—of values, norms, virtues, purposes, meanings—then of course it makes no sense to oppose it in the smallest way, or to express any but the most casual concern for any creature beyond what its utility to humanity might justify.[32] If it is a resource for humans, then and only then does it make sense to be concerned for its proper "management," although never for its existence for its own sake: that would be meaningless at best, and more likely "crankily misanthropic."

Fulford is correct—that is, is logically consistent—to dismiss "animal liberation," not because he singles out any particularly appropriate criticisms of the concept, but rather because he detects a theological impropriety in even *attempting* to elevate anything to the status of humanity. The "religion of humanity" makes such a project unthinkable. Persons advocating the cause of nonhuman beings certainly do not appear to stand in proper awe of a species whose members "exercise their imaginations, create grandly, see themselves in history, come to some understanding of the world." Fulford evokes two important points in this short quotation that relate directly to our discussion. First, he speaks with dismay at anyone putting humans on the "same level" as other beings, a situation that is only offensive, surely, if one has very low regard for that to which one is being compared. And indeed, since "sub"-human animals are commonly understood as merely behaving matter that could not possibly have purpose or value, it must surely seem "cranky" to express concern for these creatures. Second, in his evocation of the superiority of humanity, he mentions as evidence of that superiority the fact that they can "see themselves in history" (although one wonders what that says for the millennia of human activity that precede any consciousness of history). Once again, it is a central article of humanist faith that we dwell in history, that is, that we are "human-made." As creators of the cosmos, we are obviously beyond comparison with other beings: we are the only things of value, for we invent it in our activity, in our ability to exercise imagination, "create grandly," and see ourselves in history. This analysis is entirely consistent. It is the only conclusion Fulford could reach, and he does so honestly and eloquently—but without any attention to the underlying assumptions that make his diatribe inevitable.

Those assumptions, however, are extremely important to our exploration of "nature" because they are indispensable to the maintenance of the fundamental distinction "human-nature." Each of these two categories is defined by the other, in some measure: what is in nature is the not-human. Yet this distinction is compromised by the fundamental ambiguity that the term *nature* has displayed from the outset, a circumstance that has permitted the mingling of material and normative substances within a single domain. But in the humanist project of defining humanity, it has become imperative that the distinction be purified; there may be no mingling of that which is human with that which is not.[33] And since norms and values are asserted to be the coin of the free and self-willing human being, they clearly must not be attributed to nature: hence the requirement that the humanist priest perform the necessary exorcism to drain the remnants of subjectivity from the corpse of nature.

Although we have been considering the struggle to contrast humans and nature in a contemporary context, the process is far from new, and it is certainly not trivial. Our examples provide only slight embellishment to what C. S. Lewis referred to as "that great movement of internalisation, and that consequent aggrandisement of man and desiccation of the outer universe, in which the psychological history of the West has so largely consisted."[34] In that same conceptual upheaval in which we can discern the beginnings of humanism, we also see nature taking its modern form. It is to that period of transformation that we must now attend.

Part 2

The Creation of Nature

We reduce things to mere Nature in order that we may "conquer" them. We are always conquering Nature, because "Nature" is the name for what we have, to some extent, conquered.

C. S. LEWIS

3

The Purification of Nature

The term *nature,* as we have seen, is a strange one, one that almost seems to invite misunderstandings. Since its material aspect seems to give credence to whatever normative qualities are attributed to it, it provides a ready shelter for social ideals. To many, that very ambiguity constitutes a danger that demands action: a constant scouring to ensure that any normative content be reassigned to humanity, leaving only a material residue in nature.

The concern over the strict maintenance of these conceptual boundaries becomes apparent from the very beginnings of humanist writing, such as Pico's *Oration on the Dignity of Man* (1496). He has God proclaiming to Adam that "the nature of all other creatures is defined and restricted within laws which We have laid down; you, by contrast, impeded by no such restrictions, may, by your own free will, to whose custody We have assigned you, trace for yourself the lineaments of your own nature."[1] The humanist rejects the very possibility of limits on humanity and asserts that nature is strictly constrained. With that assertion goes the implicit belief that nature is thus knowable: if nature is shackled by law and cannot move, it is ours to interrogate at will.

A methodological requirement for this interrogation is apparent: one must remove any obscuring contaminants so as to obtain full access to the victim. If nature's visage is masked by a coating of projected human qualities, then obviously the way to a clear understanding is to scrape off that coating to expose the underlying reality. Of course, such a conclusion demands several assumptions: that humans do indeed have their own "content" to project; that such projection from inner to outer is possible; that there is an external, material "screen" upon which these inner phantoms may play; and that the "real" nature, the screen, can be clearly discerned when such contamination is prevented. Hence, the desire to discover certain and absolute knowledge of nature hinges upon a particular understanding of the human, as well as on a particular understanding of what it means to know something. Again, the definition and understanding of the natural is linked to that of the human.

We can better grasp the significance of such definitions if we attend to their origins. It is unnecessary (and doubtless impossible) to identify one single point in history as the decisive one in the genesis of the modern understanding of nature. One could no doubt find significant points of change across several centuries, but the Italian Renaissance of the fifteenth and sixteenth centuries seems to exemplify best the quality of that transformation.[2] And since it provides a dramatic contrast between what has become the modern understanding and all that went before it, we will follow Ernst Cassirer in concentrating on this period as our watershed to modernity.[3]

Nature and Empathy

Given our contemporary acceptance of the necessity of objectivity in the study of nature, we are apt to be easily sympathetic with those who first sought to achieve an uncontaminated vision of the world. Indeed, from our perspective, it would appear that they had even more cause than we to fear such contamination, for the society around them was more than willing to attribute nonmaterial qualities to nature. One might say that it insisted on so doing, for nature was not taken to be merely material. Furthermore, it was accepted by many that the way of knowing nature had to be through a kind of empathy,[4] which is of course "only possible if the subject and the object, the knower and the known, are of the same nature; they must be members and parts of one and the same vital complex. Every sensory perception is an act of fusion and reunification."[5] Such an equation of human and nonhuman is to our minds fundamentally erroneous, or worse: it invites the kind of projection that Barthes, on behalf of all of us, so stridently condemns. And this change of view very nearly epitomizes the quality of the transformation which we find in the Renaissance.

Before we discuss the significance of this reversal in attitude, we might consider what empathy implies: an underlying similarity between the human and the natural world. For nature to be knowable through empathy, subject and object must be fundamentally akin. It would therefore be difficult to postulate a nature that was entirely devoid of the qualities we associate with humans. Furthermore, it would be difficult to deny to nature the *variability* that humans display. Indeed, the medieval mind accepted a level of uncertainty that we would find intolerable. J. H. Randall, Jr., noted that "to the man of six hundred years ago, anything might happen in this world. Nothing was too strange or too contrary to nature for him to credit on respected authority. . . . This faith in the miraculous grows out of an attitude of mind that colored everything in the Middle Ages, from the chance event to the cosmic sweep of Providence, the desire to understand, that is, to find a meaning and a significance in things."[6] Nature's behavior, therefore, cannot confidently be predicted, for it is potentially capricious; indeed, to imply otherwise is to exclude the possibility of miracle, a blasphemous assertion.[7]

Yet even if the means to knowledge often revolves on empathy, the purpose of the medieval search for knowledge of nature is to achieve, indirectly, knowledge of the will of the Creator.[8] Nature is not simply a physical entity but a record of the will of God. This nature, encrusted with "signatures," possessed a symbolic content that was far more significant than its material content. Indeed, M.-D. Chenu suggests that it was very nearly the touchstone of this period that "all natural or historical reality possessed a *signification* which transcended its crude reality and which a certain symbolic dimension of that reality would reveal to man's mind. . . . Giving an account of things involved more than explaining them by reference to their internal causes; it involved discovering that dimension of mystery."[9] Given that understanding, it would seem foolish for a searcher of knowledge to concentrate on the lesser realm of matter—the "internal causes"—rather than on the greater realm of meaning. Or, to phrase it in Aristotelian terms, it would not do to concentrate solely on *efficient* causes to the exclusion of the important *final* causes, a contrast that helps reveal the nature of the impending change as well as the magnitude of the mutiny that was accomplished by the Renaissance revolutionaries.

Aristotle's four causes—material, formal, efficient, and final—were taken for granted at the time, but seem needlessly complex in this, the age of the single cause. Indeed, it is hard to think of some of Aristotle's categories as causes at all; hence, Toulmin and Goodfield's ingenious and helpful suggestion that we might more comfortably think of them as the four *be*causes. This is admirably illustrated by their example of a large statue composed of butter. To explain its collapse, one might say that the material cause was "because it was made of butter." The formal causes concern what the thing is in essence: "it collapsed because it was an oversized statue." The efficient cause refers to what precipitates the change: "it collapsed because the sun's warmth softened it." And the final cause is its significance or reason for being: "it collapsed as a portent to men."[10]

In considering the affairs of nature, the efficient cause would be of principal interest today. But of the others, the one with which we would be least comfortable is the final cause: the purpose *beyond* the mere physical event, which now seems an utterly unnecessary appendage to a perfectly rational explanation. But it was the final cause that was essential to the medieval mind, for it was understood that "every creature in the world is, for us, like a book and a picture and a mirror as well," and that "the entire sense-perceptible world is like a sort of book written by the finger of God."[11] Nature is God's handiwork, replete with messages that people must discern, the better to live by God's will. It is an amazing prospect, to dwell in a world in which each element is potentially meaningful, and which must be read like a book, not dismantled like a machine. It is difficult to imagine walking through a world which is actually able to inform one, and in which what is seen is never "all there is." What rich potential lies in every sight, and what great hope for understanding lies around every bend.[12]

The "finger of God" did not, however, write in clear, discursive, unidimensional prose; it wrote in symbols. It was essential that it do so, for only through a symbol can the "message" be given; the truths in question defy mere words. Symbolism was used "to give primary expression to a reality which reason could not attain and which reason, even afterwards, could not conceptualize."[13] Reason was, therefore, inadequate to the full exploration of nature, for reason could not comprehend or express that which was of greatest importance. That assertion was vociferously challenged in the revolt to which we must presently turn our attention, but it is important at this stage that we fully appreciate the medieval belief that a person must acquire knowledge of God *through* the world. In elucidating this medieval belief, Chenu observes:

> Connatural with matter, a man's intelligence had to work through matter to attain a grasp of transcendent realities, unknowable in themselves. Paradoxically, man's intelligence, in its operation, must heartily accept matter and yet must austerely pass beyond it. The movement between the two aspects of this operation was not a matter of simple psychological transference or of aesthetic interpretation; it derived from the very nature of things, which, given their sacred emanation, comprised graded representations of the inaccessible One, the Godhead. This *anagoge*, this upwards reference of things, was constituted precisely by their natural dynamism as symbols. . . . The symbol was the means by which one could approach mystery; it was homogeneous with mystery and not a simple epistemological sign more or less conventional in character.[14]

In the medieval view, simply knowing the material aspects of an object does not give access to the significant, divine aspect of being: "to stop with things themselves, whether through passion, through self-concern, or even for love of science, was to miss both true science and sanctity at the same time."[15] Symbolism could not be ignored, any more than could the principal cause of interest, the *final* cause: it was *purpose* that mattered, not simple causality. The science of Thomas Aquinas, based in large measure on the writings of Aristotle, did not share modern science's yearning for prediction and control, but rather sought a kind of understanding and contemplation that can best be described as wisdom: "a comprehension of the meaning and significance of things, above all the chief end of man, the meaning of human life and of all creation as related to it; and hence its object was that which alone gave purpose to existence, God."[16] It asked not "how" but "why."

But there was a division in the way in which symbolism was regarded, one which in some measure facilitated the transformation that occurred in the Renaissance.[17] Chenu refers to this as a profound difference, evident in the contrast between the Neoplatonism of Augustine and that of "pseudo-Dionysius" (an elusive and anonymous figure in Christian history).[18] Augustine's 'sign' was a more inward or personal phenomenon: the knower gave the sign its value beyond the material.[19] This means of course that the inner life of the individual takes precedence over any external reality. Exterior objects and events were mere stimulants, never the source of meaning. This system of belief would be compatible with the later focus on the individual, and on the human person as the sole source of value and purpose, for it makes the human the final authority.[20]

In contrast, the pseudo-Dionysian view was that "it was not the believer who gave signs their meaning; it was objective elements themselves which, before anything else and by their very nature, were so many representations, so many 'analogies.' The symbol was the true and proper expression of reality; nay more, it was through such symbolization that reality fulfilled itself."[21] One can well imagine the horror with which a modern critic, such as Roland Barthes, would view such an assertion, for here again he would see the attribution of human-generated meanings to an external source. The means by which this medieval symbolism was said to function also illustrates its contrast to modern convention, as demonstrated by Barthes. The latter's explanation of sign-creation as the juxtaposition of signifier and signified so that the latter "colors" the former into a new and meaningful sign could almost be said to be a "contamination" theory of meaning: in the juxtaposition, one factor is stained by the other until a new entity, a "passionified rose," emerges.

But of pseudo-Dionysius's "law of symbolism," Chenu notes that "the essential appeal of [this] dialectic . . . seems to have been that it bridged the apparently nontraversible gap that the mind perceived *between two realities otherwise akin;* to join these two realities within a single symbol was to put the mind into secret contact with transcendent reality.[22] As we shall see, this preoccupation with symbolism was to change quite abruptly in the Renaissance. But for the moment, let us simply acknowledge the medieval predisposition to seek empathetic knowledge of the world, and to search for signs instead of surfaces. Symbolic meaning is every bit as much a part of nature as are its physical properties, and ideas were to be sought *in the objects of nature.* Furthermore, the fact that "the cosmic symbolism of pseudo-Dionysius tended to relegate any reference to history to a place of secondary importance"[23] is also an interesting contrast to the contemporary view, and one that would seem to obstruct the path to modernity.[24] For although the Augustinian bias apparently comes to dominance in the modern notion of the individual who functions in history, at the time of the Renaissance revolution the alternate view would have commanded attention. It is not surprising, then, that those, like Galileo, who eventually chose to challenge it did so emphatically.

That the change was dramatic suggests that the critics held views so different from their predecessors that there was no middle ground between them. They are polar opposites, and their mutual repulsion creates an instantaneous and apparently unbridgeable void. Perhaps the majority of Renaissance society held views less extreme than either of these alternatives, but people comfortable in their compromise are seldom motivated to advocacy, and it is the record of the advocates that is left for us to interpret. Although we take considerable risk in focussing on a few individuals while discussing changes that involved a whole society, the fact that we all have personal experience of the polarities that exist among people makes it easier for us to appreciate the circumstances that might precipitate a transformation of societal understanding. While the question of differences among human individuals is not our central concern here, it does provide us a means of access to the larger question of societal transformation. As William James observed, the philosopher "trusts his temperament. Wanting a universe that suits it, he believes in any representation of the universe that does suit it."[25] Perhaps what happened in the Renaissance revolution is somewhat analogous. If so, examining the contrast of temperaments may help us gain some feeling for the schism that emerged.

Abstraction and Empathy

The fact that two persons, apparently of equal ability and similar background, can come to diametrically different understandings of an event has intrigued thinkers of every society and has often provoked new typologies to better describe the contrasts. The ancients spoke of "phlegmatic" and "melancholic" characters; now, in the age of popular psychology, descriptions of character seem to arise with the frequency (and ambiguity) of astrological forecasts. One typology that has stood the test of time better than most is Carl Jung's basic division of "introversion" and "extraversion." Of course, Jung compounded this polarity with a plethora of subsidiary divisions, but these need not concern us here. What makes Jung's typology useful to us in this discussion is its characterization of these contrasting approaches to the world and the parallels they exhibit with historical trends in the description of nature.

Jung contrasts the introverted and extraverted viewpoints by reference to each type's attitude to the object. The introvert encounters a world of active agents, objects that seem empowered and willful, and that therefore constitute potential obstructions to his or her own existence. Hence the introvert's effort to defuse the power of the object by withdrawing from it, creating instead an inner realm in which elements abstracted from the world are imprisoned in a conceptual edifice designed to contain them.[26] In contrast, the extravert perceives no such world of willful objects, and instead undertakes to animate the encountered world. Indeed this empathetic approach to the object is so generous that there is some risk of losing the self. But the consequence is an encounter with a world that is never foreign, hostile, or empty: it is all, in a sense, "like me." Furthermore, since the extravert's attention is directed outward to the external world, it is in its objects that truths and ideas are found. They are "discovered" *in* the objects because "the essence of empathy is the projection of subjective contents."[27] Jung summarizes these contrasting attitudes thus:

> The man with the abstracting attitude finds himself in a frighteningly animated world that seeks to overpower and smother him. He therefore withdraws into himself, in order to think up a saving formula calculated to enhance his subjective value at least to the point where he can hold his own against the influence of the object. The man with the empathetic attitude finds himself, on the contrary, in a world that needs his subjective feeling to give it life and soul. He animates it with himself, full of trust; but the other retreats mistrustfully before the daemonism of objects, and builds up a protective anti-world composed of abstractions.[28]

If we can imagine the dilemma of a person of the first type, the introvert, when encountering a reality defined by those with a more extraverted attitude, we may be able to appreciate the psychological violence of the confrontation. Not only does the world so described not make sense to the introverted mind, it is fundamentally unsettling, if not frightening. For it presents an empowered world, a potentially dangerous world, and—perhaps most important—an *alien* world. This is, of course, not the experience of the empathetic thinker, for whom the qualities of that world are his or her own; there is no question of any threat or want of meaning. But the introvert, at least in the extreme embodiment (which would surely be a small minority), could not accept this: I stress, *could not* accept this. I do not mean to imply that it is simply distasteful, but that it is fundamentally incompatible with the person's own experience and needs. And the only defense available is abstraction.

Obviously, we all use abstraction in the course of the daily practice of being human. But Jung is alerting us to a preferred *attitude,* not to the facility of abstraction. When we abstract, we "take away" or "withdraw"—we remove something from the totality.[29] Abstraction is not only a noun, as we commonly use it, meaning "the idea of something which has no independent existence" or "a thing which exists only in idea." It is also a verb: "the act or process of separating in thought, of considering a thing independently of its associations";[30] it is an action, not simply an idea divorced from reality. And it is the abstracting action of a certain element of Renaissance society that is critical in the emergence of the modern understanding of nature. The analogy is apparent: just as Jung's introvert draws away from the empathetically empowered world described by the extravert, so a particular segment of Renaissance society recoiled from the picture of the world bequeathed them by their medieval forebears. The description they rebelliously asserted was one quite unlike anything seen before. But equally important was its audacious claim to *truth.* The medievals had grand abstractions of their own, of course; indeed, they delighted in "system building." But their approach was significantly different, for they sought only to "save the appearances," to design a model that accommodated reality without internal contradiction simply for the pleasure of constructing it. That is, they approached their system much as we might a work of art, as a thing to be appreciated for its elegance or beauty, rather than as a literal imitation of reality.[31] But the approach of our Renaissance abstracters was quite different: they not only rejected the vision of their forebears, not only asserted a different description of nature, but *declared their new model to be literally true.* In so doing, they not only began the process of revisualizing nature, but also of transforming what is regarded as *knowledge.*

As we shall see, there were circumstances that favored the transmission of this new vision to other society members. Nevertheless, it is remarkable that it could be received so willingly, and while we may speculate as to why this could be so, the important point is that it was so: that the society of the Italian Renaissance was ready to be persuaded that nature was utterly unlike what they had assumed. In retrospect, it does not seem that the critics of the empathized and symbol-ridden nature of the middle ages mounted any surprisingly effective arguments against it. Rather, they appear to have managed to outlaw the very possibility of such content in nature. Remember, abstraction is "a form of mental activity that frees [a particular] content from its association with irrelevant elements";[32] hence, the first step must be to identify and dismiss those troublesome elements. Taken to extremes, this can entail mistrust of the general human perception of nature, dependent as it is on idiosyncratic and subjective properties of the person.[33] This separation of the "real nature" from the "merely human" is prerequisite to the celebration of a grand abstraction; only when the "real" can be torn free of all "irrelevant elements"—including human attributes such as color and smell, as well as meaning and purpose—can it be safely installed on the pedestal of truth. The "real" world is entirely *outside of* humanity, beyond and unlike. Hence, as Hans Jonas observes, it was asserted that "final causes have relation to the nature of man rather than to the nature of the universe—implying that no inference must be drawn from the former to the latter, which again implies a basic difference of being between the two. *This is a fundamental assumption, not so much of modern science itself as of modern metaphysics in the interest of science.*"[34] The power of the external world is neutralized by simple denial, and the new vision is sustained through the dismissal of all unwanted contents as "projection."

System

This expulsion of qualities from nature, although radical in the extreme, was justified through the assertion that only when the distorting effects of human projection are removed can we achieve an understanding of the "primary" or real properties of nature, properties that can be articulated through the perfect language of mathematics. An archetypal understanding of these geometric shapes was presumed to be lodged in the human mind, awaiting the call to service. Yet this is a faculty which few are fully able to employ, and once the use of normal perception is banished, society becomes dependent on a secular priesthood for its knowledge of nature.

Or, perhaps we should say, for its knowledge of *Nature,* Nature with a capital,* for in this process of abstraction the

*To minimize confusion, I will use this convention of capitalizing "Nature" when referring to this specific idea or system, in contrast to "nature" in the more colloquial sense of the natural world.

great domain of "everything-but" becomes even more tightly circumscribed: it becomes emphatically evident that Nature must be the "not-human." The distinction is not new, but the stridency with which it is asserted in the Renaissance certainly is. The medievals also afforded humanity a special role, but that role entailed a kind of mediation between God and nature;[35] it did not assume a strict exemption, nor did it exclude properties found in humans from the domain of nature. The distinction, which is made specific in Galileo's separation of the "primary" (real, mathematical) properties from the "secondary" (human) ones, is, as E. A. Burtt observes, "a fundamental step toward that banishing of man from the great world of nature."[36] Indeed, that is what Nature is: a world devoid of the properties we associate with humans—in short, devoid of *subjectivity.*

The search for final causes in Nature is not only futile, but invalid: there are no "purposes" there to discover. Yet humans do persist in searching for such things: apparently they must. And since they must, the "nature of humanity" is, in this view, a source of constant defilement for natural philosophy. Hence, final causes must be actively expunged along with all other "anthropomorphic" projections. Yet as Jonas notes, "It has never been argued that final cause is a far-fetched or abstruse or even 'unnatural' concept—on the contrary, nothing is more cognate to the human mind and more familiar to the basic experience of man: and this was precisely what in the new scientific attitude counted against it. Our very proneness to final explanation makes it suspect."[37]

In other words, the discernment of meaning or purpose in Nature is to be regarded as a *conceptual pollution* of reality. The discovery of inappropriate elements in nature is frequently explained as instances of "projection," a notion which Dutch psychiatrist J. H. van den Berg identifies as a means of dismissing any observation not in accord with accepted belief.[38] It is a means of disposing of "crazy ideas" before they pose a challenge to a social system. Projection is regarded as no less serious a pollution than the dumping of dioxins in drinking water. The latter threatens the purity of "the environment"; the former, the purity of "Nature." A threat to the purity of a system, any system, is an act of pollution, and the identification of an act of pollution always arouses the system's defenders. Recall Mary Douglas's explanation, that there can be no dirt without System, for dirt, by definition, is "matter out of place," and all places are determined by System. As she observes, "Uncleanness or dirt is that which must not be included if a pattern is to be maintained. To recognize this is the first step towards insight into pollution."[39] And in our understanding, in our "system" called Nature, human characteristics are out of place and constitute a serious contamination. This, as Cassirer notes, was particularly evident in the Renaissance transformation of nature. In contrast with the medieval approach of empathy and union as a means of knowing, the Renaissance abstracters sought the opposite: the removal of any trace of fellow-feeling in pursuit of an utter withdrawal from nature.*

The system that was established in the Renaissance is *explicitly* devoid of human participation, and the devotion to this ideal is shared even by nonscientists who, like Roland Barthes, would protect the realm of history by restricting access to Nature. If no human properties belong in the system of Nature, then in Douglas's terms they constitute dirt and must be expunged. All that belongs in a system of Nature is "necessity," the expression of obligatory relationships that can be described through mathematics. Galileo makes the extent of his housecleaning apparent in his famous statement about the "real" properties of nature, those which do not rely on the senses of the human being: "Hence I think that tastes, odors, colors, and so on are no more than mere names so far as the object in which we place them is concerned, and that they reside only in the consciousness. Hence if the living creatures were removed, all these qualities would be wiped away and annihilated."[40]

For Galileo, "the individual sense perception, no matter

*Cassirer claims that "the opposite form of interpretation is found in that study of nature that leads from Cusanus through Leonardo to Galileo and Kepler. It is not satisfied with the imaginistic and sensible force of the signs in which we read the spiritual structure of the universe; instead, it requires of these signs that they form a system, a thoroughly ordered whole. The *sense* of nature must not be mystically felt; it must be understood as a logical sense. And this requirement can only be fulfilled by means of mathematics. Only mathematics establishes unequivocal and necessary standards against the arbitrariness and uncertainty of opinions" (*Individual and Cosmos*, p. 54).

how intense or forceful it may be, is a mere 'name'; it neither 'says' anything nor has any objectively definite meaning. Such meaning is born only when the human mind relates the content of the perception to the basic forms of knowledge, the archetypes of which are in the mind itself. Only through this relationship and this interpretation does the book of nature become readable and comprehensible."[41] Galileo's is, of course, the view of knowledge we have come to accept as truth, even though neither we nor its official practitioners, whom we call scientists, are likely to have given much thought to its fundamental assumption of separation. Yet all cooperate in the attempt to explain life in terms of nonlife.*

If the living creatures were removed, contamination would cease and the "real" Nature would stand revealed. It is a staggering undertaking, the effective expulsion of all human, or indeed living, properties from the system that is Nature.[42]

Anthropomorphism

Exclusion is, of course, essential to the establishment of any system. "In chasing dirt, . . . we are governed not by anxiety to escape disease, but are positively re-ordering our environment, making it conform to an idea."[43] But what is unusual in this system is the category constituting dirt: "our own felt aliveness."[44] Indeed, the system which we know as Nature excludes just this, from the outset: "life" is dirt. Hans Jonas identified this phenomenon when he observed that prior to the Renaissance, life was the norm and death was the anomaly to be explained. After the Renaissance, death is the norm and life the anomaly to be explained—or, at least, to be set aside so as not to confuse the issue.[45] Certainly, there was no edict to the effect that life was not part of nature. The expulsion

*Jonas remarks that "the question regarding life now poses itself thus: How does the organism stand in the total context already defined, how is this special order or function of it reducible to its general laws—how, in short, is life reducible to nonlife? To reduce life to the lifeless is nothing else than to resolve the particular into the general, the complex into the simple, and the apparent exception into the accepted rule. Precisely this is the task set to modern biological science by the goal of 'science' as such. The degree of approximation to this goal is the measure of its success; and the unresolved remainder left at any time denotes its provisional limit, to be advanced in the next move" (*The Phenomenon of Life*, pp. 11–12).

was subtle and incremental, beginning with the elimination of human sense-perception and continuing with the gradual elimination of anything that might be considered "subjective," that is, derived from living *subjects.* And the sin we commit in attempting to attribute such properties to Nature is "anthropomorphism."

By the dictionary definition, anthropomorphism is "an interpretation of what is not human or personal in terms of human or personal characteristics."[46] It is the ascribing of human characteristics to nonhuman things. Originally, this term was linked to a form of heresy, that of seeing God in the form of man, of giving God human characteristics. Anthropomorphism still is heretical, but the direction of the transgression has been reversed. In the original sense, it was an affront to God to be made human; now, in the age of secular humanism, it is an affront to humans to have their form given to what are called "subhumans." The reason given for resisting this offense is that it is a false comparison—which, in one sense, it certainly is. Taxonomically, only one species has human form. It is no longer the *form* that we are concerned about, however, but the *characteristics* or *qualities* of humans.

There is a more subtle form of anthropomorphism, as in our earlier illustration in which the various "camps" ascribed differing human qualities or *ideals* to the functioning of nature—competitiveness, cooperativeness, dominance, aggressiveness, and so forth. Claiming to find these properties in nature allowed each group to legitimate its own prescription for the ideal society. But since we cannot know with certainty the motivations of another society of creatures, we are surely practicing a kind of anthropomorphism in claiming to do so—perhaps what we would call a "cultural anthropomorphism" rather than an "emotional" or a "physical" anthropomorphism. When we speak of things like competition, we are dealing with something rather different than the original anthropomorphic heresy. Instead of attributing a human form, emotion, or sensation to something, we are attributing a human *explanation* to the nonhuman world. We see two mammals bumping each other or making loud noises, but what we report is often not a simple description, but a conclusion: the animals are exhibiting "aggression," or perhaps "competition."[47] Through this form of anthropomorphism we are apparently able to impute certain social interpretations to nature, thus perpetrating the kind of mythmaking challenged by Barthes.

Anthropomorphism, then, takes at least three different forms: the attribution of *human form,* the attribution of *human characteristics,* and the attribution of *cultural abstractions.* It is the latter variety that Barthes has understandably identified as a danger. The first two seem, on first glance at least, to be somewhat more obvious and perhaps easier to grapple with. They also seem to be less dangerous, although still indicative of the boundary transgression which we so abhor.

There are some inconsistencies, however. For example, although we may dismiss the embodied sensations that make the world seem "like me," the attribution of human physical form to others now seems strangely inoffensive, much less so than was the attribution of human form to God. Indeed, one could reasonably argue that all medical research is based on the tacitly accepted legitimacy of this form of anthropomorphism. We are entirely willing to assume that many other animals have human-form, that is, that their tissues and organs function just as ours do. Because of that, we are able to perform experiments on them and extrapolate to ourselves. If the analogy were utterly false, we would not have modern—that is, experimental—medicine at all. It would appear that there is a tacit acceptance of the legitimacy of a strictly physical kind of anthropomorphism: other creatures *do* have human-form, at least on the tissue and organ level. The acceptability of this, however, probably indicates our willingness to abandon our bodies to Nature: that is, we think of our physical bodies as merely the material containers of the real human person. Since such containers are from the realm of Nature, it is quite understandable, and permissible, that they share characteristics with other parts of Nature, for example, with other animal bodies. Bodies are no longer truly *human* at all (a belief which may partly explain both the societal tolerance of abortion, in which no "person" seems in evidence, only a fetal "body," and our rejection of euthanasia in those cases in which we feel a person is clearly present).[48] The rise of biotechnology can only reinforce this attitude, and it may be that the division of human self from human body is so complete as to make this form of anthropomorphism inconsequential. The ban on transference of human properties to nature has not been lifted; the body, however, has been disinherited and now remains a mere physical adornment of the self, a piece of organic jewelry which the "body-building" industry promises to cut and polish.

Yet while we may be able to accommodate such modest transgressions, Barthes's concern is by no means uncommon, and the epistemological policing of Nature is very much the concern of the modern system of education in the West. This education, in which we invest so much money and faith, ostensibly to make ourselves more "competitive" with other industrialized societies, is in essence an intellectual manicure that will scrape the dirt of human perception from the understanding of nature and so maintain the purity of that external realm. It is also, of course, a means of maintaining the conceptual segregation of humanity and nature. E. A. Burtt provides a dramatic encapsulation of the emergence of the new understanding of nature:

> The gloriously romantic universe of Dante and Milton, that set no bounds to the imagination of man as it played over space and time, had now been swept away. Space was identified with the realm of geometry, time with the continuity of number. The world that people had thought themselves living in—a world rich with colour and sound, redolent with fragrance, filled with gladness, love, and beauty, speaking everywhere of purposive harmony and creative ideals—was crowded now into minute corners in the brains of scattered organic beings. The really important world outside was a world hard, cold, colourless, silent, and dead; a world of quantity, a world of mathematically computable motions in mechanical regularity. The world of qualities as immediately perceived by man became just a curious and quite minor effect of that infinite machine beyond.[49]

In summary, we observe a dramatic contrast in the approach to the natural world between those living before the Renaissance revolution and those after it. The empathized world of the medievals is dismissed as impermissible in the new abstracted system called Nature. That system, relying as it does on the strict limitation of the permitted contents of Nature, requires a deliberate cleansing, followed by an active defense against any recontamination through "projection." With the exclusion of all "human" qualities, and of the older means of knowing nature, the locus of knowledge shifts from the world to the human—that is, to the "real" human, the inner being whence the true form of Nature derives.

Nature is, in effect, redefined, and the means of knowing nature are utterly transformed. Perhaps most important, the very concept of *knowing* is subtly altered, for knowledge is no longer an intuition of shared properties or of meanings beyond appearances, but a deduction of systemic rules from a new blueprint of Nature. Nature was, we may recall, originally taken to mean "everything." Next, everything-but-something, perhaps everything but God. But as humanity appropriated qualities formerly shared, it too came to be sharply delineated from the domain of nature. When we finally remove *anything* that might reveal traces of innate purpose or will, anything beyond inert, rule-obeying matter, then life itself is effectively excluded from Nature. Of course, animals are still associated with Nature, but only through the conceit of defining *them* as behaving matter, instinct-bound functionaries of the Laws of Nature. Rather than seek norms of behavior in nature, humans must now dictate norms *for Nature.* In this sense, Nature is *mythologized* from its inception: its contents are established through historical decree. There can be no exceptions: Nature is the realm of necessity, and there is no room for self-willed beings with purposes of their own. The conceptual purity of the domain of Nature is a condition for the security of the realm of Humanity.

4

From *nature* to *Nature*

The analogy we considered in the last chapter, between the approaches to knowing apparent in Jung's typology of human individuals and those apparent in different periods of human thought, may well be just that: an analogy. We need not posit any shift in the actual composition of human temperaments to appreciate the parallel between the two. For our purposes, what is significant is that there was, over a period of time, a rejection of the accepted way of knowing. Whereas before the Renaissance it had been considered essential that some form of empathy be employed to acquire knowledge of nature, it is now essential that empathy not be employed. And whereas it had been believed that knowledge is to be acquired through nature and the meanings discernible therein, it is now believed that knowledge of Nature results from applying human reason to discern the necessary form of Nature. This discernible form is not posited as a mere "model": on the contrary, it is taken to be the true shape of nature, freed from distracting detail and perceptual distortion. The question remains: How was it possible for this latter assertion to take priority over custom and experience as the true description of nature?

In retrospect, it might be possible to persuade ourselves that the evident logic of scientific revelation prevailed, and that the society of the Italian Renaissance was eager to escape past errors and embrace the new. Yet even if that society was less segmented into jargon-bound specialties than is our modern world, clearly the arguments of a Galileo could only be addressed to the literate few. The rest of the people, all of whom are free to report to each other what nature "is," might well remain entirely unchallenged. How then might so bold a vision as a new Nature be put to the larger society? The answer, I suspect, is that no overt argument or logical demonstration was necessary, in the initial stages: it would suffice to show the world what nature is "really like." As will become apparent, the means existed to do just that through the medium of art.*

Galileo had a predecessor in his quest, and it is the amphibious genius of Leonardo da Vinci that guided Galileo toward the delineation of the "laws of Nature."[1] As both artist and scientist—and as one who had no interest in differentiating the two—Leonardo was able not only to participate in the search for the hidden forms of nature, but to depict his discovery in visual terms. "In a manner that is characteristic and determinative of the total intellectual picture of the Renaissance," says Ernst Cassirer, "the logic of mathematics goes hand in hand with the *theory of art.* Only out of this union, out of this alliance, does the new concept of 'necessity' of nature emerge. Mathematics and art now agree upon the same fundamental requirement: the requirement of 'form.' "[2] As an artist, Leonardo was centrally concerned with the question of

*I do not mean to suggest, of course, that visual art is the only medium through which alternative views of nature might be transmitted. Rather, I select this one as emblematic of the whole realm of creative discourse, and one which is especially useful for our purposes because of its accessibility. For a discussion of the wider question of the role of "creative speakers" in social change, see Richard Rorty, *Contingency, Irony, and Solidarity* (Cambridge: Cambridge University Press, 1989).

proportion, which was of course an instance of applied mathematics as well as an aesthetic device. "Thus the idea of measurement becomes the connecting link, joining the natural scientist to the artist who creates a second 'nature.' "[3]

Leonardo was not content simply to be a superior technician, applying geometric formulas to create appropriate illusions. He wanted to be the *creator,* the one who actually establishes the form of reality. An "ideal 'blueprint' must precede the mind's *working* upon nature" and it is just this blueprint which the creator must devise. Leonardo "emphatically states that nature is full of 'rational principles' *that have not yet been part of experience. . . .* Galileo follows the same path. Though he considered himself a champion of experience, he nevertheless emphasized that the mind can only create true, *necessary* knowledge *by its own principles.*"[4] Notice the order of priorities: it is all very well to attend to experience (direct experience, of course, not to alleged "signatures" or other immaterial properties), but when a choice must be made, the *inner experience* takes precedence: the reasons are more "true" than the objects. Just as the medieval investigators of nature sought purposes or final causes, those of the Renaissance sought reasons—or rather, *provided* the reasons of Nature through their creative endeavor. Nature was thus subsumed to the human enterprise, a puppet of the rational forces discernible through the human mind. The colors discernible through that mind are inadmissible after Galileo; the reasons discernible through it are, however, considered fundamental.

Much of our admiration for Leonardo arises from his new concept of the *necessity* of nature—that nature is a rule-bound, law-abiding entity, following forms that are knowable to the human mind. It has only one "correct" form, and it is that one form that the scientist or artist must discover and define. "Reason is the immanent, unbreakable law governing nature. Sense, sensation, or the immediate feeling for life can no longer serve as the means by which we assimilate nature and discover her secret. Only thought proves to be truly equal to nature; only 'the principle of sufficient reason,' which Leonardo considers a principle of mathematical explanation."[5] Once we assume that there is *of necessity* a single way of nature, a single pattern by which it is bound, then it is indeed knowable—not by experience, in the colloquial sense, but by reason alone. Hence, humans need not seek guidance in the world; rather, they detect the inevitable in it, and they become not receivers of an external wisdom but revealers of an innate and necessary form. Both nature and humanity are cast in new roles, with new properties and expectations. Science and art are merely different ways of giving form to nature. "Be it as artist or as thinker, the genius finds the necessity in nature."[6]

The genius finds the necessity in nature: this sums up the base assumption which can never again be ignored, and which implicitly supports all that follows. Yet necessity is also part of the *abstraction* that Leonardo discovered within himself, an abstraction that he was able to offer *verbally and visually* for others to embrace.[7] Necessity becomes, therefore, the standard of nature. Nature, though explicitly nonhuman, is *ours:* we do not so much read the "book of nature," as Galileo desired, as write it. It is a human artifact, and like the traffic light, its only purpose is provided by the use we give it.[8] With such complete freedom, humans are released to "exercise their imaginations, create grandly, see themselves in history, come to some understanding of the world," as Robert Fulford asserts. Indeed it is doubtful they could do so without having made this transformation of nature, for as Cassirer observes, "reflection on human freedom, on man's original, creative force, requires as its complement and its confirmation the concept of the immanent 'necessity' of the natural object."[9] For the humanist concept of "Human" to exist, we must first invent Nature: our freedom rests on the bondage of nature to the "Laws" which we prescribe.*

*C. S. Lewis makes the interesting observation that although we think of the medieval views of nature as highly anthropomorphic, our own understanding may be even more so. He notes that in the medieval view, "there were four grades of terrestrial reality: mere existence (as in stones), existence with growth (as in vegetables), existence and growth with sensation (as in beasts), and all these with reason (as in men). Stones, by definition, could not literally strive or desire.

"If we could ask the medieval scientist, 'Why, then, do you talk as if they did,' he might (for he was always a dialectician) report with the

The Substitution

The wonderful, geometrically precise paintings of Leonardo were, of course, matched and amplified by other painters of the time, and his vision of Nature was duly embellished in words and numbers by those who followed in the wake of Galileo. Certainly geometry had been a subject of fascination long before the Italian Renaissance, but it had largely been understood as a description of *ideal* forms, not actual ones. One of Galileo's great feats was to extend this to *all* phenomena, by the simple assumption that in each object lies hidden just such a perfect form as is described in geometry, a form that can inevitably be discerned and revealed through the application of mathematics. That is, freed of the irregular encumbrances that conceal them, the *true* form of objects can be discerned and described. With that belief came the will to interpret not just ideal forms in this way, but all forms. Thus began, as Husserl notes in his *Crisis of European Science,* "the surreptitious substitution of the mathematically substructed world of idealities for the only real world, the one that is actually given through perception, that is ever experienced and experienceable—our everyday life-world." Galileo was not troubled by the fact that in actual experience forms are never exact, only approximate. He was sufficiently confident that the world *should* be geometrically perfect to assume that it *must* be so, and to set about describing it as if it were, while elaborating a method to facilitate the transcription.[10]

Once the substitution is accomplished, then in effect the new idealized or abstract system becomes, by definition, Nature. Henceforth, if we ask about "nature," we need look only this far, not to that which lies in the experience of us all.

---

counter-question, 'But do you intend your language about *laws* and *obedience* any more literally than I intend mine about *kindly enclying*? Do you really believe that a falling stone is aware of a directive issued to it by some legislator and feels either a moral or a prudential obligation to conform? We should then have to admit that both ways of expressing the facts are metaphorical. The odd thing is that ours is the more anthropomorphic of the two. To talk as if inanimate bodies had a homing instinct is to bring them no nearer us than the pigeons; to talk as if they could 'obey laws' is to treat them like man and even like citizens (*Discarded Image,* pp. 93–94).

"Mathematics and mathematical science as a garb of ideas, or the garb of symbols of the symbolic mathematical theories, encompasses everything which, for scientists and the educated generally, *represents* the life-world, *dresses it up* as 'objectively actual and true' nature. It is through the garb of ideas that we take for *true being* what is actually a *method.*"[11] We would do well to note Martin Heidegger's claim that the concept of the mathematical derives from the Greek *ta mathēmata,* which means "what can be learned and thus, at the same time, what can be taught"—in other words, "the *mathēmata* are the things insofar as we take cognizance of them as what we already know them to be in advance"—"the mathematical is 'about' things which we already know. Therefore we do not first get it out of things, but, in a certain way, we bring it already with us."[12] It is certain knowledge, for it is a priori knowledge and not something discerned in a mysterious "outer" world: we bring the answer with us when we pose the question. So too in art: "In proportion, the mind re-discovers itself and its own rule."[13]

While claiming to have found a way to discern reality accurately, the emerging system of thought tended to deprecate ordinary perception in favor of an idealized version. Leonardo speaks of experience, but when he "refers to experience, it is to discover there the eternal and unchangeable order of reason. His true object is not experience itself but the rational principles, the *ragioni,* that are hidden and, so to speak, incorporated in experience."[14] It is a commonplace that the Renaissance figures broke from the blinders of scholasticism to observe the world directly, but in fact the observation was itself filtered by new expectations. In effect, there was a rejection of prior assumptions about the means to knowledge in favor of a different assumption: truth through abstraction. It is not enough to see or touch, and certainly not to court a "fellow-feeling" of nature; one must seek the hidden truth, the real essence of nature, which is discernible only through reason, and more precisely, through mathematical reason. It is as if, by shaking the autumn tree, the leaves can be made to drift away to reveal a trunk composed of the number one, and branches of twos and threes. The picture of nature that will emerge is not that of common experience, but one that no one would even suspect until introduced to the new model; it is, so to speak, a non-experienced reality which is introduced as the one true Nature.

The Portrayal of Nature

Again, it seems improbable that so novel an understanding of nature could have been successfully insinuated through the use of reason and mathematics alone, even if it is upon these that it depends. To be accepted, it would have to be revealed persuasively. And the tools for this task were providentially available. In light of Ernst Cassirer's observation that the refiguring of nature occurred both in science and in art, it is hardly surprising that a genius whose interests so clearly spanned both realms should figure prominently in the revelation of Nature. Indeed, Cassirer further suggests that "it was artistic 'vision' that first championed the rights of scientific abstraction and paved the way for it."[15] At the very least, Leonardo is emblematic of the kind of agent that would be necessary to effect a change of this magnitude.

We are not inclined to ascribe such potency to artists today, perhaps because their role is so very different now. Moreover, we tend to find detailed "theoretical" or "scientific" statements very persuasive: their sheer density seems to command respect. Yet their effect may be quite small compared with the more mundane expressions of daily life. Each of us makes explicit our postulations about the world each time we say, "I see a tree over there." The objects that we see, and report having seen, are revelations of our understanding of reality. Indeed, as J. H. van den Berg observed, "If we want to understand man's existence, we must listen to the language of objects."[16] Our assumptions are most boldly revealed when we don't even realize we are using them, and for us the realm of nature is populated by literal, not symbolic, objects.

When we simply reveal what we see, what objects populate our world, we state quite clearly our expectations of the world. Since we do so in the belief that what we report is "really" what we see, however, we are unlikely to be charitable toward the visions of others. For instance, we are prone to look on cave paintings, or those of any pre-Renaissance artist, as representations of a flawed illustrator. But it may be more useful to regard these as remnants of another reality, glimpses into the way the world "really" was when those people made their images, than to regard them as failed copies of the One True Nature.[17] Indeed, the fact that we do find all pre-perspective painting to be wanting is evidence of our own habits of thought and sight, for it is by means of that technique that our new world is revealed.[18] This change in focus is readily apparent in the paintings of the Renaissance, and most particularly in those of Leonardo. Van den Berg suggests that the new nature is first revealed in Leonardo's masterpiece, the *Mona Lisa.*

The landscape depicted behind the *Mona Lisa* was a novel one. It is an area of the painting which we see as "plain nature," untainted by codified significance. Rilke, whom van den Berg credits with the first realization that a new nature appears with this painting, concisely characterizes the changing nature of artistic representation of, and human relationship to, the natural world.

> No one has painted a landscape which is so entirely landscape and yet so much confession and the painter's own voice as is the depth of background behind the Madonna Lisa. It is as if all that is human were present in her infinitely quiet portrait, but as if everything else, all in front of man and beyond him, were in this mysterious complex of hills, trees, bridges, sky, and water. This landscape is not the picture of an impression, not a man's view of quiescent things; it is Nature which came into existence, a world which grew and was as foreign to man as the untrodden forest of an undiscovered island.[19]

This is a remarkable transformation, from the sympathetic and meaning-rich foreground of the medieval to the distant, ordered landscape of the Renaissance. Rilke says that "men only began to understand Nature when they no longer understood it":[20] when nature was no longer akin to humans, when it was freed of symbolic meaning and "anthropomorphic projection," then it could be discovered and its true form revealed. But the fact that Nature is unlike us meant that we were finally and utterly alone.[21]

The consequence of this was monumental, not simply for nature, but for humanity. According to Rilke, "The artist immersed himself in the great quietness of things, he felt how their existence was passed within laws, without expectancy and without impatience. And the animals passed quietly amongst them and suffered, like them, the day and the night and were full of laws."[22] The creatures of Nature were indeed now full of laws, and of very little else. The medieval person had lived in a world of potential surprise and continuous significance. It was rich with *content.* But with Leonardo, the world congeals to a necessary *form.* "Viewing it only through the medium of art, Leonardo does not see nature devoid of form but rather as the very realm of perfect and complete form itself. Necessity does, indeed, reign as the bond and the eternal rule of nature—but this is not the necessity of mere matter; it is the necessity of pure proportion, intimately related to the mind."[23] Furthermore, it is understood that the perceived form *must* be the attribution of reality, for "man can only be certain that the sense world *has* form and shape if he continually *gives* it form. . . . 'Oh investigator of things,' says Leonardo, 'do not praise yourself for your knowledge of things brought forth by nature in its normal course; rather enjoy knowing the aim and the end of those things designed by your mind.' "[24]

Remember, Leonardo regarded the artist as "Lord and Creator," a person who is able to constitute an ideal world and from whom "'abstraction' and 'vision' collaborate intimately."[25]

> So if he desires valleys or wishes to discover vast tracts of land from mountain peaks and look at the sea on the distant horizon beyond them, it is in his power; and so if he wants to look up to the high mountains from low valleys or from high mountains towards the deep valleys and the coastline. *In fact, whatever exists in the universe either potentially or actually or in the imagination, he has first in his mind and then in his hands,* and these [images] are of such excellence, that they present the same proportioned harmony to a single glance as belongs to the things themselves.[26]

The artist presents an ideal world that can be taken in by the viewer in "a single glance" through a "proportioned harmony." Leonardo gives the viewer *a whole landscape* as a visual object. As E. H. Gombrich notes, Leonardo's landscapes are conceptual, owing less to the painter's eye than to the imagination.[27] Leonardo directs his viewer's eye to the beauty of his *abstraction,* and away from a world "contaminated" through *empathy.*

Technique and Transformation

The contrast between abstraction and empathy in art has been discussed periodically since Wilhelm Worringer's book of that name was published in 1908. It was revived most recently by the psychiatrist Anthony Storr in his discussion of solitude, in which he observes that the abstracting tendency of an artist constitutes an attempt to "create order and regularity in the face of a world in which he felt himself to be at the mercy of the unpredictable forces of Nature." As noted earlier, the empathetic interpretation sees no such danger, and indeed in extreme forms leads to the possibility of losing oneself *in* the object. As a defense, abstraction is particularly effective when it is self-contained and complete. Storr notes that "geometric form, on the other hand, represented an abstract regularity not found in Nature," and as such is especially appropriate. "Abstraction, then, is connected with self-preservation; with the Adlerian, introverted need to establish distance from the object, independence and, where possible, control."[28] There is, perhaps, no greater relief for the introverted mind than the conviction that the external world is bound by a systematic order and is therefore susceptible to prediction and control.

This, of course, is consistent with our earlier discussion, in which these contrasting tendencies were used to illustrate the forces that might tend to facilitate transformations in the understanding of nature. But Storr resists any temptation to distinguish between art and science in this regard:

> Thus, abstraction is linked with detachment from the potentially dangerous object, with safety, and with a sense of personal integrity and power. This is also the kind of satisfaction which the scientist experiences in his encoun-

> ters with Nature. A new hypothesis leading to a law which will predict events originates from perceived regularities, from the ability of the scientist to detach himself, his own subjective feelings, from whatever phenomena he is studying, and, when proven, gives an enhanced power over Nature.[29]

It is appropriate to recall that this is just the kind of transformation that Cassirer notes in the Renaissance transformation, that "neither art nor mathematics can allow the subject to dissolve in the object and the object to dissolve in the subject. Only by maintaining a *distance* between the two can we possibly have a sphere for the aesthetic image and a sphere for logical-mathematical thoughts."[30] In the Renaissance, we see both scientist and artist creating a purified *alternative* to the external world, a new "Nature" whose genesis is within the individual and which is subsequently used to make sense of external objects.

Worringer was fully aware that behind abstract or "transcendental" art lies an urge to control: hence there seems "only *one* possibility of happiness, that of creating a world beyond appearance, an absolute, in which it may rest from the agony of the relative. Only where the deceptions of appearance and the efflorescent caprice of the organic have been silenced, does redemption wait." He further concludes that all such art "sets out with the aim of de-organicising the organic, i.e., of translating the mutable and conditional into values of unconditional necessity."[31] But curiously, he does not associate this tendency with Renaissance art. Rather, he presumes that the need which was once satisfied through abstraction in art is now satisfied through science, leaving art to become purely empathetic. But in our terms, this interpretation is exactly reversed: Leonardo's art was certainly not geometric in the sense that tiled mosaics or patterned blankets can be, but it was nevertheless an attempt to render reality as if it were inevitably enslaved to geometric laws, those we associate with perspective and proportion. Leonardo's world looks "real" only after we are able to read it as a system of regularities. Moreover, it was clearly an attempt to abstract from the total context of chaotic nature an ordered subset that can "stand for" the whole in his system of thought. Worringer's description of the urge to abstraction could very nearly have been a description of Leonardo:

> Tormented by the entangled inter-relationship and flux of the phenomena of the outer world, such [artists] were dominated by an immense need for tranquillity. The happiness they sought from art did not consist in the possibility of projecting themselves into the things of the outer world, of enjoying themselves in them, but in the possibility of taking the individual thing of the external world out of its arbitrariness and seeming fortuitousness, of eternalising it by approximation to abstract forms and, in this manner, of finding a point of tranquillity and a refuge from appearances. Their most powerful urge was, so to speak, to wrest the object of the external world out of its natural context, out of the unending flux of being, to purify it of all its dependence upon life, i.e. of everything about it that was arbitrary, to render it necessary and irrefragable, to approximate it to its *absolute* value.[32]

To take "the individual thing of the external out of its arbitrariness" is exactly what Cassirer detects in Leonardo's quest: "transforming empirical accidentality into orderly necessity."[33] His chief means of realizing this vision of purified form had been provided by the upsurge of interest in linear perspective, which is intimately intertwined with the emergence of the new Nature. Linear perspective was certainly not unknown in earlier times, but in the Italian Renaissance, and particularly in the Florence of about 1425, it rises to prominence. The return of the humanist Alberti may have been the precipitant of significant change, for he put forth a popular and detailed treatise on the subject. But from the outset, as Samuel Edgerton, Jr., points out in his *The Renaissance Rediscovery of Linear Perspective,* "this new Quattrocento mode of representation was based on the assumption that visual space is ordered a priori by an abstract, uniform system of linear coordinates." This assumption permits the artist to locate himself in one position and describe an objective field to a similarly placed viewer, his partner in the perception of a grand and geometric abstraction. In contrast to the "visual world" of earlier art, which represented the scene as experienced and felt, this "visual field"—in James J. Gibson's term—is simply what is perceived when the viewer can immobilize both his body and his intellect to record the optic impression only. Instead of the diffusive image that emerges through constant scanning, peripheral vision, and emotional rendering, the fixated eye reveals the singular effect of an immobilized lens. To those introduced to the technique at this time, "linear perspective . . . with its dependence on optical principles, seemed to symbolize a harmonious relationship between mathcmatical tidiness and nothing less than God's will."[34]

Yet Alberti's technique was not intended, as we might assume, to better render a specific place more accurately. Rather, "it provided a purely abstract realm which the viewer would discern as a world of order: . . . *real* space in the sense that it functioned according to the immutable laws of God."[35] This is the same sentiment evident in Johannes Kepler's assertion that "geometry is unique and eternal, a reflection of the mind of God. That men are able to participate in it is one of the reasons why man is an image of God."[36] And since, as Alberti and his collaborators assumed, the world functions best when ordered, when obeying mathematical laws, it follows that the world depicted as obeying those laws must be "more true." Such was the character of the "tool" that was presented to Leonardo, a tool ready-made to the execution of his design.[37]

Edgerton observes that the use of such a technique to produce an image of uniform, ordered space would not have been impossible for medieval artists, but it would have been contrary to their intentions. In the light of the Aristotelian concept of heterogeneous space, space differentiated into places appropriate to the differing qualities of objects, the prospect of representing space as homogeneous is essentially unthinkable.[38] The question is not what is technically possible, but what is conceptually appropriate, and uniform space was utterly inappropriate to the depiction of the medieval world. By the same token, however, the fact that it *was* appropriate to the representation of the Renaissance world indicates how fundamental a transformation is in the works: a new space, a new world, has emerged. Edgerton observes that "finally, in the fifteenth century, there emerged mathematically ordered 'systematic space,' infinite, homogeneous, and isotropic, making possible the advent of linear perspective."[39]

Whether the advent of perspective follows or precedes the change in the notion of the world is debatable, but linear perspective evidently provided the means to a presentation of a new vision of the world—a vision that could also be disseminated widely, thanks to the availability of printing. Without this joint availability of perspective and printing in the Renaissance, there might well not have been an upsurge of science and technology.[40] Yet the artistic revelation of this new concept of space and reality may have been the fundamental prerequisite to the development of science: Leonardo's "vision of nature proved to be a methodologically necessary transition point, for it was artistic 'vision' that first championed the rights of scientific abstraction and paved the way for it."[41] Leonardo had to show us what this new world looked like before we could begin to investigate it; he made it real.[42] Indeed, he made it the definition of the real.

Certainly, in the early stages of the application of perspective to painting, the viewer would need to be geometrically "literate," to be able to "read" such a painting. The educated public of the Italian Renaissance apparently was particularly well prepared in this regard: the principal component in their educational system was mathematics, including geometry and, most especially, proportion. Although it would have been taught principally for commercial purposes, Michael Baxandall suggests that "the literate public had these same geometric skills to look at pictures with: it was a medium in which they were equipped to make discriminations, and the painters knew this."[43] The whole social environment was conducive to the acceptance of the primacy of a geometric world, revealed or hidden, as the basis of reality.

Leonardo's audience, then, was amply prepared to interpret the new paintings, and thus to become the occupants of his new Nature. With nature shorn of symbol, the leaves of grass stand forth in their empty simplicity. This too is significant: the very possibility of literal objects—what we would today simply call "objects"—was unthinkable until this transformation, for objects had never been "merely" objects until this time.[44] Before there could be a literal object, all metaphoric or figurative content had to be dismissed. Leonardo gives us a vision of reality which makes this possible, and thus sets the stage for the increasing attention to, and trust in, the literal surfaces of the natural world.

5

The Literal Landscape

In establishing a new "landscape of reason" Leonardo was, in a sense, following in the tradition of Italian art: the illumination of a general truth at the expense of individual variations. In emphasizing the underlying rationality of Nature, he had to overlook the individual and irregular particulars that obscure the essential regularity—a tendency that has been noted even in the scientific work of Galileo.[1] Yet despite the Italian preference, there is also a wealth of detailed information in the paintings of Leonardo and his contemporaries. In this paradoxical combination of the general and the particular, Leonardo anticipates the next stage in the evolution of "Nature."

Creating the Real

One of the principal social functions of medieval and Renaissance art was to educate: an essential story or meaning was to be transmitted to the viewer. The picture *denotes* something: it has a message, and that message constitutes the reason for the painting's existence. Even if a painting does not illuminate a traditional religious theme, it is still expected to contain a core of meaning, not simply an efflorescence of detail. How, then, are we to account for the addition of detail apparently superfluous to the communication of the central theme?

In his innovative critique, *Vision and Painting: The Logic of the Gaze,* Norman Bryson has commented on this question, and in particular on a semiological interpretation that seems to provide an explanation. While the theme denoted in a painting is probably well known to the viewing audience, the impression of *veracity* can be considerably increased through the addition of detail which the viewer "discovers" independently. That is, although the central theme is articulated through standard iconographic conventions, that which is *beyond* the theme lacks an interpretive context, unless the viewer can provide it independently. This in turn makes the viewer a participant in the authentication of the painting, and imparts a sense of heightened realism to the entire scene.[2] "If at the centre of the image an enthroned Madonna is encircled by angels, then it is towards the landscape at the sides that the realist gaze will tend, finding in a frozen pool, in a peasant carrying wood, in a dog scurrying across a field, more of the truth of the Middle Ages than in the triumph of its Church; not because it lacks faith or trusts only to a low-plane reality, but because the form in which it perceives truth is in the absence of bias: given sufficient 'honesties' to deflect attention towards the insignificant, its credulity is without bounds."[3]

The suggestion that truth is to be found in the periphery, in the "unauthored" areas of depiction, is especially interesting. Remember, the prerequisite to knowledge of the new Nature is the relinquishing of all notions of human involvement in it: there are no more "signatures" in Nature. Once it is accepted that the reality of Nature is devoid of patently human involvement, then what will be discerned as most truthful is that which appears free of human content and authorship. The "externalities" of the painting, the inconsequential detail of leaf or flower, bear no impress of artistic message or intent: they appear as simple reports of what is "really there," regardless of the whims of humanity. That is their power: their *ap-*

parent independence of human intent. They have not been insinuated by any authority, and they can be read without aid of translation. The viewer interprets the painting in the context provided by his or her own life, rather than relying on the context of iconographic authority.[4]

The apparent absence of human interference, which contributes to the sense of authenticity, leads to a further development, the scouring of the evidence of authorship not only from the scene depicted but even from the medium of its expression. To the extent possible, not even evidence of brush strokes is allowed to betray the human touch. Indeed, in the later and more sophisticated uses of linear perspective, not even the body of the viewer is permitted to intrude: only the "gaze" discerns this "objective" scene: "Western painting is predicated on *the disavowal of deictic reference,* on the disappearance of the body as the site of the image; and this twice over: for the painter, and for the viewing subject."[5]

The painting Bryson selects to illustrate this development is Vermeer's *The Artist in His Studio,* in which we find the body of the painter in the scene being depicted. The body is represented in the scene as, in effect, a natural object, a thing among things.[6] The viewer, the "real" person, is a *dis*embodied eye. In Bryson's encapsulation, the

> first geological age of perspective, the epoch of the vanishing point, is the transformation of the subject into object: like the camera, the painting of perspective clears away the diffuse, non-localised nebula of imaginary definitions and substitutes a definition from the outside. In its final form . . . the only position for the viewing subject proposed and assumed by the image will be that of the Gaze, a transcendent point of vision that has discarded the body of labour and exists only as a disembodied *punctum.*[7]

The painter in question is a seventeenth-century Dutch artist who demonstrates a significant transformation of the use of linear perspective. In her discussion of Dutch art of that time, Svetlana Alpers underlines its divergence from the ideal of the Italian Renaissance, and suggests that the contrast between the two styles illustrates differing ways of picturing the world. In the Italian tradition, we find "the picture considered as an object in the world, a framed window to which we bring our eyes," while in Dutch painting we see "the picture taking the place of the eye with the frame and our location thus left undefined"[8]—as a "disembodied *punctum,*" perhaps.[9] The significant point is that exact compliance with a geometric system of representation is no longer the primary objective. The *system*—the expression of the reasonable structure of nature that subtly becomes the denotative "core"—is of less importance than the literal phenomena recorded by the alien eye.

Leonardo was to elaborate a new standard for nature thanks to the fortuitous arrival of a graphic tool appropriate to his purposes. It seems unlikely that anything short of linear perspective could have facilitated this transformation. Once presented in elegant simplicity, authenticated by the connotative participation of the viewer, Leonardo's Nature becomes the norm to which experience must conform. As Cassirer reminds us, "The decisive point in Leonardo's thought is precisely that a dualism between the abstract and the concrete, between 'reason' and 'experience,' can no longer exist."[10] Experience must conform to reason, and his landscape is embodied reason.

Yet there is a certain ambiguity in Leonardo's vision, which he seems unable to resolve. On one hand, he is very much the system-builder who is intent on revealing the rational structure of Nature. Yet at the same time, his interests were so broad that he often reveals a *non*-systemic interest in the world: he is clearly fascinated by the visual surfaces around him. It is unlikely that he could have expressed his interest in simple appearances on a formal scale in a society that still looked for "general human traits and general truths" in each painting.[11] Indeed, such indulgences were openly condemned, as in the following comments attributed to Michelangelo.

> In Flanders they paint with a view to external exactness or such things as may cheer you and of which you cannot speak ill, as for example saints and prophets. They paint stuffs and masonry, the green grass of the fields, and shadow of trees, and rivers and bridges, which they call landscapes, with many figures on this side and many figures on that. And all this, though it pleases some persons,

> is done without reason or art, without symmetry or proportion, without skillful choice or boldness, and, finally, without substance or vigour.[12]

Remember that to the Renaissance artist, proportion was of great importance. For a painter to ignore the rational principles of composition would suggest incompetence, and to create paintings which lacked theme or message would seem pointless: where was the narrative that gave purpose to the painting? Landscape detail was meant to be ancillary to the central theme, not a substitute for it. Bryson observes that "in realism, periphery counts over centre."[13] In effect, these northern landscapes had *only* periphery: they left out the center.

The Dutch painters, however, did not favor center over periphery, any more than they did the general over the particular: quite the contrary. Indeed, they were suspicious of attempts to discern general truths, and shared Francis Bacon's view that "to resolve nature into abstractions is less to our purpose than to dissect her into parts."[14] They would forgo the grand overview in deference to fascinating particulars. The Dutch, like the English, came to believe that knowledge was to be had on the surface, through accurate description of the experientially given, rather than through introspection.

Leonardo, ever enigmatic, seems sympathetic to both Italian and northern preferences, and cannot give himself fully to either.[15] Despite his fascination with appearances, Leonardo fears that "the painter who draws merely by practice and by eye, without reason, is like a mirror which copies every thing placed in front of it, without being conscious of their existence."[16] Given his belief in the artist as the active creator of a world, he cannot accept the role of a passive reporter.[17] Thus, even though he shares with the Dutch a fascination with visual detail, he could not accept the role of *mirror* of nature the way they apparently could. The questions that arise, then, are why they could and did, and what bearing this has on the description of nature.

Seeing Is Believing

In *The Art of Describing,* Svetlana Alpers suggests that Dutch society was predominantly visual, in contrast to the textual orientation of the Italians. "In Holland," she says, "the visual culture was central to the life of the society. One might say that the eye was a central means of self-representation and visual experience a central mode of self-consciousness.[18]

"Seeing is believing" was nowhere a more appropriate adage. This is apparent not only in art, but in all kinds of pastimes, one of the most notable being their fascination with microscopes. It was this passion that led to contact with the Royal Society in England, whose members shared this enthusiasm. However, the English concern with objects and surfaces found a different mode of expression: experimental science. "Both Dutch descriptive painting and English empiricist science involved a perceptual metaphor for knowledge. . . . The basis for certain knowledge was to be nature witnessed. The craft of the painter, and the art of the experimentalist, was, therefore, to make representations that reliably imitated the act of unmediated seeing."[19]

Notice that it must be *unmediated* seeing, with the interference of the observer kept to an absolute minimum. In their study of early English empiricism, Steven Shapin and Simon Schaffer note that "the solidity and permanence of matters of fact reside in the absence of human agency in their coming to be."[20] It is this predisposition in the treatment of the periphery in Renaissance paintings that Bryson commented on as well. But this predisposition is taken to an altogether new level in northern art, which concentrates on surfaces unmediated by any semiological or iconographic system of interpretation. Increasingly, the assumption is made that facts can be accurately discerned by the passive mirroring of reality.*

We have the ground laid, therefore, for a different conception of what constitutes a fact, and of how to obtain a true understanding of nature. The details of the external world, which Leonardo enjoyed but could not make his major focus,

*The notion of "mirroring" reality is discussed at length in Richard Rorty's *Philosophy and the Mirror of Nature.* In it, he discusses the transformation from the "hylomorphic conception of knowledge," in which "knowledge is not the possession of accurate *representations* of an object but rather the subject's becoming *identical* with the object," to the Cartesian notion that the representations are in the mind and therefore available for contemplation. Hence "the Inner Eye surveys these representations hoping to find some mark which will testify to their fidelity" (p. 45).

now come to be seen as the very means to knowledge. And just as the Italians had used extraneous detail to give a sense of veracity to their paintings, the English experimentalists came to embellish their reports with detailed illustrations, which seemed to imply that the verbal descriptions were similarly fastidious.[21]

If we look at a Renaissance painting and then at a Dutch landscape, the difference may seem slight. Both, after all, appear "realistic." Both use the technique of linear perspective, in one form or another. But we must not underestimate the significance of a change in subject matter, however subtle it may seem. The significant transformation in the Renaissance was from a nature of symbols and sensibilities to a Nature of certainty and reason. It was the systemic *structure* of Nature that was created, a structure that defined rational limits to Nature's behavior. And while empiricists certainly must begin with a notion of an ordered and consistent world, they attend to the visible surfaces of that world,[22] rather than to its rational substructure. Systems are all very well, they would say: but they are human-made, and can be unmade. The objects of nature, however, are simply *there:* they are *factual.* And they may be known without the necessity of theory and conjecture, by the simple expedient of looking with intent.

I do not mean to suggest that the Italian preoccupation with the reasons of Nature was to be utterly displaced: far from it. The rationalist philosophies of continental Europe pay ample tribute to that heritage, and still foster debate with Anglo-American empiricists over the correct means of knowing nature.[23] But for the most part, the English mode prevails in both modern science and popular assumptions.[24]

The Dutch artists, who were as taken with the method of induction proposed by Francis Bacon[25] as were the English, provide a visual representation of the world the empiricists were intent on exploring. The latter were by no means unchallenged in this pursuit, even within England: the controversy between Thomas Hobbes and Robert Boyle is illustrative of this.[26] But for our purposes the most interesting part of this transformation is the attention to literal detail, to the *surfaces* rather than to the systems of Nature. The Dutch artists of the time are sometimes described as achieving an "almost photographic" accuracy; they are even accused by some of having relied on the "camera obscura" to obtain some of that accuracy. That they *knew* of the camera obscura was more important than that they used it, however, for it came to represent the very model of seeing and ultimately of knowing.[27] That is, the idea of perception as an act in which an entire image is projected passively onto the retina—some actually referred to the image as a "picture"—suggests a model for how we know nature.[28] By exposing ourselves, via our sense organs, *to* nature, we are able to receive an accurate impression. And since it comes to us directly, without apparent intervention, that impression is obviously a *true* one. A person is, in fact, a kind of camera: the only trick is to ensure that one is a *good* camera, which may be accomplished by the precise techniques of Dutch attention to detail or English experimental observation.[29]

The idea of the human as analogous to the camera in its mode of acquiring knowledge accentuates the significance of both sight and visual surfaces. Sight, in its apparently passive reception of information, is the most trustworthy of our sources of knowledge—which leads us to concentrate increasingly on visual information, and even to contrive methods by which phenomena normally open to our other senses may be "translated" into visual form.[30] This gradual transition to a single mode of sensory access biases the *ratio* of sensation, and leads to a preponderantly visual understanding of reality. Perhaps, as Hans Jonas suggests, "the mind has gone where vision pointed."[31] But certainly, attention to visual surfaces remains our most trusted means to knowledge.

The acceptance of the "camera" model of human perception also leads to a camera model of human knowledge: that our only means of access to the true images of the world is to become polished *mirrors* of that world; we must improve our *representations* of reality, both in the mind and on the canvas, if we are to improve our knowledge of it. Toward this end, there arises a fascination with representation per se, and with the question of how the mind is able to assess the relative merits of each representation. Hence, the whole idea of a "theory of knowledge" emerges in response to the concern that our "inner representations" be accurate.[32]

Speaking of Nature

A truthful representation is, clearly, one that corresponds exactly with the surfaces of nature. The acceptance of the literal truth of correspondence, however, is not limited to the sphere of the visual. If anything, it is even more apparent in our use of words. There are, of course, a great many theories of language development, and one could scarcely hope for widespread agreement on this topic. But for our purposes, the brief description provided by Northrop Frye in his *The Great Code* will suffice to illustrate this point. Essentially, Frye argues that the language favored by earlier societies tended to be predominantly metaphorical: hence their tendency to attend to the similarities of things, while we are inclined to focus on distinctions. A second phase emerges, however, when people begin to put this for that—that is, when "words are 'put for' thoughts, and are the outward expressions of an inner reality. But this reality is not merely 'inside.' Thoughts indicate the existence of a transcendent order 'above,' which only thinking can communicate with."[33] This is just what Leonardo does in the establishment of his system of Nature: he uses the language of reason, and assumes that the model he creates is in fact a revelation of a transcendent order given by God, the Great Geometer.

Frye's third phase of language, however, is most interesting, for it seems to run parallel to the rise of landscape painting. It begins in the sixteenth century, and achieves dominance in the eighteenth. "In English literature, it begins theoretically with Francis Bacon, and effectively with Locke." Again, in this third phase, there is the essential separation of subject and object, and an exposure of the subject to the objective world which is "the order of nature," while "deductive procedures are increasingly subordinated to a primary inductive and fact-gathering process." This is, again, the "camera" model of knowledge acquisition, and one which trusts simple fact-gathering to deductions from elaborate intellectual systems.

> Hence this approach treats language as primarily descriptive of an objective natural order. The idea to be achieved by words is framed on the model of truth by correspondence. A verbal structure is set up beside what it describes,

> and is called "true" if it seems to provide a satisfactory correspondence to it. The criterion of truth is related to the external source of the description rather than to the inner consistency of the argument.[34]

It is not the argument, the rational exposition, which is the locus of truth, but the literal correspondence to external surfaces.[35] Not surprisingly, this understanding of truth enraged William Blake, who rejected the idea of passively achieving knowledge of the world without the engagement of the imagination, which could reveal that which is *more* real, so to speak—that which transcends the limitations of sensation or reasoning, and which constitutes wisdom rather than mere knowledge.[36]

Blake was not unique in this idea, and the position he advocated has certainly not disappeared, despite the predominance of the opposing view.[37] It may be useful to point out that this, too, seems analogous to the conflict of temperaments described in Chapter 3. Recall that we examined Jung's description of introversion and extraversion as being "distinguished by their attitude to the object." That is, the expectations of the objects, the external world, are fundamentally different in the two approaches, one experiencing an essentially animate world of potential danger, the other a neutral world to be animated by the person him- or herself. This has an obvious bearing on one's approach to nature, and to the understanding of what constitutes knowledge *of* nature.

Remember, the knowledge sought by the introvert is essentially an internal image or *idea* of nature, which provides an explanatory system and a sense of order and control. Jung claims that the introvert "struggles against any dependence on the object, he repels all its influences, and even fears it. So much the more is he dependent on the idea, which shields him from external reality and gives him the feeling of inner freedom."[38] It is a *rationalist's* approach.[39] In contrast, the extravert is an *empiricist* who looks to the outer world for knowledge. Hence, "anything but reductive thinking is simply out of the question, since for him general concepts are mere derivatives from experience. He recognizes no 'rational concepts,' no *a priori* ideas, because his passive, apperceptive thinking is oriented by sense impressions. As a result of this attitude, the object is always emphasized: *it* is the agent prompting him to insights."[40]

This contrast seems similar to the one we see in the transformation in nature-description from that of the Italian Renaissance to that of northern art and experimental science. Rather than attend to the *idea* intrinsic to nature, to the reasonable structure through which all else can be explained, the northern approach is to attend to the literal surface of *objects*, and to find there the truth of the world. The latter looks outside to seek the answer, the former looks within to discern the truth beneath mere appearances.* Jung concludes: "The two types are opposed in a remarkable way: the one shapes the material out of his own unconscious idea and thus comes to experience; the other lets himself be guided by the material which contains his unconscious projection and thus comes to the idea."[41]

Again, it matters not that the actual players in the Renaissance tended to introversion, or those of the Royal Society to extraversion: the illustration simply dramatizes the curious oscillation of intent in the understanding of nature. In observing developments from the medieval period through to the present, we seem to see a change from an extraverted attitude to objects, then to introverted, and back at last to extraverted. Visually, we see a change from an attention to detail—"the Middle Ages are unrivalled, till we reach quite modern times, in the sheer foreground fact, the 'close-up'"[42]—to the harmonious "world-as-a-whole" of Renaissance landscape, and back again to a fascination with fragmented detail.[43] A similar tendency is apparent in language: Frye observes that in the current phase of language use "we return to a direct relation between the order of nature and the order of words, as in the metaphorical phase, but with a sharp and consistent distinction between the two."[44] But even though we again attend directly to objects, the attitude that informs our interrogation of nature is vastly different. The objects of our medieval an-

* Alpers suggests that the Dutch painters shared Bacon's belief that "to resolve nature into abstractions is less to our purpose than to dissect her into parts," and that they are willing to sacrifice "the selection of a single, prime, or privileged view ('abstraction' in Bacon's terms) that is empowered to summarize knowledge" (Alpers, *Art of Describing*, p. 91).

cestors had a built-in significance; ours quite emphatically do not.[45]

Leonardo's notion of nature-as-necessity, as a realm dominated by rational principles that can be discerned and described, is a comforting image for one threatened by an unpredictable and animate world because it implies the possibility of control. Yet strangely, the actual *quest* for control arises in the later tradition, with its belief that the way to know the world is through "trying it out." This was essentially Bacon's recommendation, but it contained within it the explicit imperative that this was to be done with the intent of *conquering* nature. Hence Jonas's conclusion that "a new vision of nature, not only of knowledge, is implied in Bacon's insistence that 'the mind may exercise over the nature of things the authority which properly belongs to it.' The nature of things is left with no dignity of its own. All dignity belongs to man: what commands no reverence can be commanded, and all things are for use."[46]

Moreover, Bacon's method contained an inherent technological potential, for it "yields directions for works because it first catches nature 'at work.'"[47] Its practical intentions color its observations, and even its concept of nature. Moreover, the entire empiricist enterprise presents us with a different understanding of what constitutes matters of *fact.* One of the conditions of true descriptions of nature is that they be visible to anyone, not confined to an intellectual elite. It is a democratization of knowledge. It was virtually essential that the ordinary person seem to participate in this, for the assertion that "plain practical folk" could verify and thereby legitimate the observations of experimental scientists was regarded as an important virtue. We don't look to the individual genius, but to the whole social consensus: a fact is true *because we all can see it.* Hence, as Jonas says, "the aristocracy of form is replaced by the democracy of matter."[48]

Owen Barfield offered a useful parable to illustrate the conflict between the two approaches. He considered a group of people, all occupants of an automobile called the Universe. Dispute arose between two groups as to how the vehicle worked, with one group attending to nonvisibles, such as internal combustion, the other to the apparent exteriors. The second group dismissed talk of internal combustion as meaningless abstraction, and insisted that the way to find out how the thing worked was by experiment, by the pushing and pulling of levers. Eventually the second group "began to maintain that the first kind of knowledge was an illusion based on a misunderstanding of language. Pushing, pulling and seeing what happens, they said, are not a means to knowledge; they *are* knowledge."[49]

Pushing and pulling things that are readily apparent in the "real world" seems the only possible route to knowledge, if one first dismisses the other method's approach as meaningless. If this kind of self-censorship can be sustained, the questions asked of nature will all be of a kind, and so too the objects thus discerned. And if, further, the *language* that is judged suitable to discuss nature is also of one kind—that which relies on literal correspondence to surfaces—then of course it is impossible to even speak seriously of any "content" of nature, metaphoric or otherwise. It simply doesn't make sense, given the tacit adoption of these conventions of discovery.

We are fully aware, and are constantly reminded, of the benefits that accrue to those who possess practical knowledge. And we are assured that greater acquisition and application of such knowledge—that is, more pushing and pulling of nature's levers—will in due course alleviate the contemporary "environmental crisis." But the fact that we constantly reassure ourselves of this may have less to do with our faith in the assertion (though it is certainly that, too) than with the fact that our very choice of language precludes the consideration of other possibilities. One can scarcely hope to be taken seriously when using, for example, metaphoric language to prescribe a solution to this crisis—such language is merely subjective, as we all know, and utterly impractical. And indeed it *is* impractical in this sense, for practicality is virtually defined as "pushing and pulling," which can be discussed comfortably only in descriptive language.

The Diminishment of Nature

The creation of the word "nature" engendered an apparent dualism in the world: all is nature or not-nature. And since the "not-nature" has come to mean essentially "humanity," our relationship to nature has seemed equivocal. At the very least, this raises doubt as to how one can even *know* this alien entity with any certainty. For the medievals, of course, "to know an object means to negate the distance between it and consciousness; it means, in a certain sense, to become *one* with the object."[50] But as we have seen, this changes with the Renaissance, when there is in effect a withdrawal from the objects of nature and a denial of the propriety of such empathy. The suspicion of any "secondary" properties—things detectable only through human senses and sensibilities—resulted in a virtual definition of truth as the *absence* of human involvement. And when the pendulum swings back from abstraction to empathy once more, it is empathy of a different kind, with a different object. The northern empiricist does not encounter objects through fellow-feeling, but through the eyes of a stranger. As Barfield observed, the empirical scientist "treats nature as an invading army treats an occupied country, mixing as little as possible with the inhabitants."[51] He no longer uses the language of symbolism, or the language of similitude: his words are exclusively descriptive and avowedly neutral.

Moreover, the objects of nature are vastly diminished, from receptacles of meaning to empty images for inspection. I mentioned earlier that in the development of the empirical method, there was considerable stress laid on the importance of the observations being evident to anyone. Indeed, it would be fair to say that it was *necessary* that "everyone" see these things, because there is now no other authority to vouch for them. In the past, the "authority" of nature was presumed to come from God: we saw it in the messages He chose to present. With the systematizers of the Renaissance, the authority shifts to what we might call the denotative core of nature: the innate "necessity" that ensures the adherence of the world to human-discerned laws. But as interest shifts from similitude to difference, and from general principles to particulars, nature is no longer "verified" by any overarching abstraction, but is instead authenticated by the general observance of humanity: nature is as it is because we can all see it. We, collectively, become the guarantors of the reality of nature. The great amorphous body of humanity becomes the new authority, and its perceptual voting power will override the testimony of individual genius.[52] Hence Frye concludes that "in Blake the criterion or standard of reality is the genius; in Locke it is the mediocrity. If Locke can get a majority vote on the sun, a consensus of normal minds based on the lower limit of normalcy, he can eliminate the idiot who goes below this and the visionary who rises above it as equally irrelevant. This leaves him with a communal perception of the sun in which the individual units are identical, all reassuring one another that they see the same thing; that their minds are uniform and their eyes interchangeable."[53] Moreover, the basis of authority in third-phase or descriptive language is also social consensus: "Hence the modern use of language has been driven increasingly to define the objective reality of the world, on the assumption that 'objective' means real, because it allows of such a consensus, and that 'subjective' means unreal because it does not."[54]

A similar observation was made in a rather different context by the psychiatrist van den Berg, who claimed that "we believe in each other—and we say that there is no other belief. We believe in what we see—and we say there is nothing beyond it."[55] But the consequence of this is a deliberate *impoverishment* of individual perception, because in order to maintain the social consensus, any deviant perception must be marginalized or eliminated. In his view, this is a major reason for the theory of projection which, though never explained or justified, is habitually employed to dispose of contrasting observations. It is apparently a theory "invented to guarantee an identity of the world for everybody, or, perhaps more accurately, to put a stop to an ever-increasing subjectivity of the world."[56] If our sole source of authority is our social consensus—"they"—then that must be protected against subjective experiences that might erode its foundations. We must insist on a world of bare surfaces, despite the consequences: "Anxious to avoid a new Babel, we would rather reduce things to an extreme poverty":[57] the lowest common nature.[58]

If we contrast the rich and heterogeneous world that was the experience of the medieval with our strictly sanitized collection of empirical objects, we can appreciate the price paid for our deference to social consensus as the sole legitimator of reality. But there has been a parallel shift in the concept of the individual, or rather perhaps an amplification of a tendency that can be observed from the beginning of the Renaissance. As nature becomes, by decree, to seem ever more barren, the individual *inner* human self becomes the postulated repository for all the qualities extracted from the world.[59] The increasingly strict division between human and nature provides a sense of secure separateness and assures that the only apparent path to knowledge of nature will be to gaze across the gulf at the visible surfaces of otherness. But perhaps the most apparent feature of the world has become the gulf itself, which constitutes a kind of moat that appears to protect while actually confining: we may be less besieged by otherness than imprisoned by self-worship.

The acceptance of such division has been prerequisite to the reciprocal development of our concepts of humanity and nature. But strangely enough, it has also become the source of a profound unease. For daily, we see signs of erosion, and we feel a vague sense of an impending in-filling which may disturb our protected domain.

6

The Fragile Division

To conclude our overview of the emergence of the modern conception of nature, we must attend to a topic that has been implicit in most of our discussion so far: the human/nature dualism. This idea seems almost too obvious to need examination, but like most such obvious ideas, it conceals many of our protected assumptions. Considering this topic, therefore, may permit us to draw together the discussion of this section and smooth the way to our final deliberations.

Dualism

Consider the development of our concept of nature as it has emerged thus far: we learned from C. S. Lewis that, although it seemed merely a matter of historical curiosity that the Greeks should have come up with a label for "everything," and then modified it to indicate "everything but" some special exception, the consequences of that labeling have been enormous. It is fair to say that before the word was invented, there *was* no nature.[1]

That is not, of course, to suggest that there were not the entities and phenomena we now attribute to nature, but rather to say that people were not conscious of there being any such entity as "nature." For nature is, before all else, a category, a conceptual container that permits the user to conceive of a single, discernible "thing." As we noted earlier, a parallel would be a fish discovering the concept "ocean." Before that, there is only water, the medium through which the animal swims without awareness of its existence, and certainly without a concept of the whole body of water being collectively containable in a concept called "ocean." But once "ocean" is created, it is possible to discuss the whole as a single entity, an object distinct from the fish.[2] And once thus possessed, it is then *given over* to this creature to hold as its own, so to speak, and to contemplate action toward.

Once, we were powerless to challenge nature. Although we certainly had an effect on the earth, we could not contemplate action toward nature per se. But all of creation became our object when the Greeks were able to capture it in a word-cage. Earlier humans could only conceive of a single world, of which they too were part. And since humans are clearly animate, conscious, and purposeful, it would make little sense to deny the existence of those properties in the rest of the world—in nature. Indeed, one could not do so without denying them in oneself. But the moment the special or *dangerous* sense of nature, as Lewis calls it, came into being, all was changed. For once there could be a separate category from all else, it became conceptually possible for "everything"—nature—to be *unlike* us. The consequences of this could not be fully realized as long as the means of knowing nature were presumed to require some empathetic identification with it. But when the path to knowledge became that of the stranger, the detached observer, there was not only the possibility but the requirement that nature be utterly devoid of human qualities.

The removal of kinship not only entails a re-assortment of worldly properties into the domains of humanity and nature, but places humans, as the beings capable of reason, in charge of that process: it gives us license to adjudicate the contents and behavior of nature. But we can succeed in this only if the categorical boundaries are strictly maintained, with no blurring of distinctions. This requires vigilance and a knowledge of the proper sanitary procedures—in short, "education." And although the grand project of boundary maintenance has been remarkably successful, it has never been entirely secure. Indeed, it is often the intensity of our territorial defense that reveals the exact location of those boundaries.

The Erosion of Dualism

Hans Jonas emphasized the dramatic shift, particularly in the Renaissance, from the fundamental assumption that the world is alive and that death is the anomaly to the assumption that death is the norm and life the anomaly. But this shift could not have been accomplished instantaneously, for the prospect is deeply unsettling. The mediating agent, the sugar that helped us swallow this new medicine, was dualism, which "is the link that historically mediated between the two extremes which so far we have opposed to each other unhistorically: it was indeed the vehicle of the movement which carried the mind of man from the vitalistic monism of early times to the materialistic monism of our own . . . ; and it is difficult to imagine how the one could have been reached from the other without this gigantic 'detour.' "[3]

Dualism *protected* us from the realization of the implications of "materialistic monism," the nihilistic understanding of the *lifelessness* of nature. We could, in time, come to regard other organic beings as "behaving matter," for we ourselves seemed protected from such reduction through the great wall of dualism. All qualities, all indications of vitality, are harbored in the realm of humanity. The meaningful part of reality is all in us: "the very possibility of the notion of an 'inanimate universe' emerged as the counterpart to the increasingly exclusive stress laid on the *human* soul, on its inner life and its incommensurability with anything in nature."[4] This is not to suggest that dualism is some kind of bizarre plot hatched to facilitate a more fundamental revolution in thought, for the distinctions it works with are grounded in our experience. But as a belief, or as an explanatory mechanism, it did permit us to go a very long way toward the material, de-animated vision of nature—until the specter of our own ingestion arose. For this is the transition of which Jonas speaks and to which we now must attend: from a "vitalistic monism" through dualism to the final "materialistic monism."

To illustrate the dilemma that arises during this transition, Dennis Lee considers the study of neurobiology.[5] The study of the nervous system, and in particular the brain, is obviously a topic for scientific investigation. The brain is a material organ whose electrochemical properties can be investigated, and that investigation in turn may inform us about the behavioral functioning of the organism. But when the behaving organism happens to be human, a dilemma emerges. The basic premise is that the phenomena to be investigated are "matters of objective fact. And the valuative properties which had been thought to inhere in them are shown *not* to be part of the objective universe. Only subjective consciousness can sustain them." Again, our basic premise: no "secondary" or subjective qualities are to be inferred in the nature of Nature. But how then are we to investigate things which are taken to *be* essentially qualitative, such as consciousness? All the study of the brain can reveal is cortical synapses, which appear to be all the brain "really" is. "The conclusion this necessitates," he says, "is as startling today as were similar conclusions about God or beauty a century ago. 'Consciousness' is merely an epiphenomenon, a subjective interpretation we have mistakenly projected onto the brain, but which has finally floated clear of its object. And any attempt to wedge it back into the still blank spaces of research will be as futile as earlier attempts to cram God into the gaps in evolutionary theory."[6]

That which is not clearly material, objective in the modern sense, must be an artifact of some sort, an "epiphenomenon." But when that epiphenomenon is consciousness, our very means of access to the "real" world, the consequences of this assumption are troubling to say the least. For as Lee concludes, "The real kicker is that with this development we no longer have access to a subject which can support, or create, the after-presence of consciousness. That is how previous values were accommodated. But now it is precisely 'the subject' which has gone up in a puff of objectification." We have examined the brain and found nobody home. We, the source of observation and of valuing, the very center of purpose-making and judgment, and our very being as subjects, are victims of the annihilating beam of objectification: we are not "real." The brain is a minefield of exploding synapses, but never a mind. "From within the liberal cosmology, there is only one conclusion open to consciousness: consciousness is dead."[7]

Interestingly, C. S. Lewis comes to a similar conclusion in his study of the change between medieval and later understandings of nature.

> In this great change there is something to be won and something lost. I take it to be part and parcel of the same great process of Internalisation which has turned *genius* from an attendant *daemon* into a quality of the mind. Always, century by century, item after item is transferred from the object's side of the account to the subject's. And now, in some extreme forms of Behaviourism, the subject himself is discounted as merely subjective; we only think that we think. Having eaten up everything else, he eats himself up too. And where we "go from that" is a dark question.[8]

Dark indeed. If "mind" is to become a disreputable concept, perhaps "humanity" itself must also disappear as a significant distinction. We have humanity, in contrast with nature, as a place in which values, thought, spirituality, and novelty could dwell. But when we began studying nature, we found it full of human *bodies*—what to do with them? Study them as we do all else, as matter, and shift the "values" to their obvious place of final refuge, the conscious mind. But the investigation revealed no mind: only a material brain. Once the entity "humanity" was laid open to examination, one statistical object among the many to be explored, the protection afforded by the category dissolved. It was ransacked and explained, and the assault on the brain now threatens to prove once and for all that the knowing subject, the very source of both the concept of nature and its investigation, is a mere epiphenomenon.

Hence, we move again from the original vitalistic monism, by degrees, through a convenient dualism to the final encounter with the new materialistic monism. We have, in effect, been consumed by our own creation, absorbed into our contrasting category. We created an abstraction so powerful that it could even contain—or deny—ourselves. At first, nature was ours, our domesticated category of regulated otherness.

Now we are nature's, one kind of object among all the others awaiting final explanation.

In the "brain" example, it appeared that the subject was lost. We know that the brain is our "factualized" being—it is the "objective" part of our behavioral existence. But there is nowhere in that brain, no part or section, labeled "self" or even "consciousness." *We,* the habitués of the special realm of "humanity," are nowhere to be found in our own bodies. And yet, we cannot discount the evidence of the body revealed through scientific questioning, for it is our one accepted means of access to the truth. Since nature is, by definition, the realm of necessity, and since science is our way of questioning nature, then we are surely not in any position to doubt its results. To do so would also be to question our assumptions of the "reality" of nature—to question the fundamental assumptions upon which the dualism is based. This is a dilemma: we cannot mistrust science, even when it proves we do not exist.

The Modern Monism

Now the final twist becomes apparent. Even though Nature is, in the first instance, a creature of history, that is, of human conception, it audaciously attempts to make history a subcategory of Nature. Once we accept, through the study of Nature, that all life is organically related, organically the same through the linkage of evolution, then humanity is literally a part of Nature. Not figuratively, not poetically, but literally an object like other natural objects. This is accepted—not appreciated, but tolerated. We cannot reject it without exposing the fiction at the core of dualism. And as Jonas points out, if humanity is "just" a part of nature, then what sense does it make to suppose that nature may not have properties similar to our own? What is the justification for the ban on anthropomorphism? And if there is none, what is the basis of the assumption that knowledge of the real has been, or could be, achieved through the exclusion of those properties known to exist in humans?[9] Once again, the dualism vanishes, and history becomes a curious subset of material nature, simply a bizarre instance of behavioral complexity that has arisen through the eternal pressures of natural selection.

We would appear to be hoist of our own petard: we have rigorously defended a dualism that permits us to think of ourselves as fundamentally unlike anything else on the planet, if not in the universe, and to establish a concept of Nature through which to characterize all else. Yet we have also used that concept as the virtual definition of that which is knowable, and we have had no reason to shun it in the examination of ourselves. When we find ourselves about to be subsumed in our own concept, our own realm of the "not-us," we have little ground for rejecting the results. The only way to get off our own dissecting table is to admit the fiction. That is, if we want to prevent the realm of humanity or history becoming a subcategory of Nature, we are going to have to admit to ourselves that Nature is in fact a subcategory of Humanity or history—that we are, after all, the authors of the system we call Nature. And moreover, that we are the authors of the dualism that facilitates the existence of humans and nature as separate and qualitatively distinct entities. We are going to have to admit our own role in the constitution of reality, which in turn means admitting something quite fundamental about the nature of our knowing.[10]

This is not to suggest that the whole house of cards is going to fall down on us tomorrow. It is unlikely that we shall ever wake to newspaper headlines proclaiming, "Dualism Disproven: Nature to Fall!" But we do see evidence of unease that bespeaks some fundamental uncertainty. We see it, for instance, in the debates surrounding intelligence tests that imply innate—that is, material—differences among equal and free-willing individuals. And perhaps most dramatically, we see it in the public response to "sociobiology," the field of study that attempts to find biological (material, natural, given) roots to our social behavior. This not only implies that what we take to be free will is in fact biologically determined, it also appears to place limits on what we are actually able to do—and that, of course, is a direct contradiction of Pico's claim that he could "do whatever I will to do." This belief, shared entirely by Leonardo and by subsequent humanists (which means essentially all of Western society), provides the very rationale for all we do, and provides the unshakable faith in "progress."[11] Leonardo believed that "the soul of man is the center of all, and thus has the force of all, and in its penetration into the secrets of nature through its art may rightfully be called the grandson of God."[12] Yet now we learn through the study of the very Nature instigated by Leonardo that there *is* no soul, only a determinate body with the limitations inherent to organic form. Our reaction, of course, is that this must be rejected, for not only is it destructive of the modern logos, it is tantamount to heresy. It is not surprising, therefore, that the scientists who dare reveal these secrets find their reception less than warm. Yet so thoroughly persuaded are they of the necessary dominance of the "Nature" monism that they seem unprepared for the response. The audacity they display in daring to treat us as if we were objects in the realm of extended matter is fundamentally unsettling—theirs is a categorical heresy, a challenge to the way the world is. It is pollution in the larger sense: one simply doesn't do that to humans. Nor does one suggest that there is anything deterministic about them: Nature is determined, but not Humanity. That distinction is prerequisite to the maintenance of the sense of human uniqueness and superiority.

Despite the centrality of this distinction, it places us in a dilemma. This dualism constitutes "the most momentous phase in the history of thought." Yet in the "postdualistic" situation we now find ourselves approaching, there appear to be only two possibilities, two vying "monisms": modern materialism and modern idealism, which "both presuppose the ontological polarization which dualism has generated."[13] Even though the dualism itself may no longer seem viable, the alternatives were nevertheless crafted in its presence and bear the stamp of its existence: each attempts an explanation of the whole from its own "pole," its own location in the dualistic setting. As we have seen, one of these polar-monisms is dominant today: "Materialism is the real ontology of our world since the Renaissance, the real heir to dualism."[14] It is also a troubled and unsatisfying one, when its consequences are squarely faced, as we saw in Lee's neurophysiological example: consciousness becomes a mere epiphenomenon of the electrochemical functioning of the brain. The attempt to explain all by reference to either of the polar-monisms appears to lead to doomed attempts to explain the whole by reducing it to a part, with the usual consequences.

Despite the problems, this dualism is firmly established and securely buried among our recessive assumptions. Thus we find ourselves in a rather schizoid state, accepting the system called "Nature" while resisting attempts to explain ourselves by it. But we must acknowledge how necessary dualism is to our efforts to maintain the strictures of humanism. Unless we have this absolute separation, we cannot claim unique qualities that justify our domination of the earth. If we are "nothing but" animals, then our claim to a hereditary monarchy is undermined. But of course, the affront we take at the comparison is also a result of the abstraction: being "nothing but" animals is only offensive if one has a derogatory view of animals, which of course our understanding of Nature requires. They are mere machinations of established and irrefutable laws of form and motion. Clearly, there is little likelihood of a "return" to the vitalistic idealism of the past, nor, in the face of our increasing reliance on both science and technology with their implicit allegiance to the concept of Nature-as-necessity, is there any likelihood of deflecting the takeover bids of the materialistic monism. The dualism that has defended us thus far seems far too fragile to withstand the assault.

Maintaining the Fiction

Joseph Campbell relates the story of a conversation he overheard while sitting at a lunch counter. A mother and son were eating beside him, and the boy remarked that his friend had written a paper on the evolution of humankind but had been chastised by their teacher, who claimed that he was wrong and that Adam and Eve were our ancestors. When the mother agreed with the teacher, the child responded, "Yes, I know, but this was a *scientific* paper." The mother was not to be placated by that proviso and exclaimed, "Oh, those scientists! Those are only theories." "Yes, I know," said the boy. "But they have been factualized: they found the bones."[15]

This instance nicely illustrates the dilemma we have been considering, and indeed that we all face daily. We exist amidst an accepted dualism: there is Nature, the domain of science, and there is history, the domain of human action. Nature can be known through scientific explanation, but should human things be "factualized?" To do so is indeed to draw humanity into the domain of Nature and to explain ourselves as the products of "natural law." But to refuse is to reject the acknowledged path to the truth, for Nature is *historically* established as the domain of the knowable, the certain, the absolute. If you want to know the truth, you use the scientific method of investigation. If you want to know the truth about humanity, you make it into an object of scientific investigation as well, a knowable object. But if you do so, you reverse the roles of the two domains, so that Nature, initially a convenient subcategory in the realm of human affairs, becomes the dominant category and absorbs all, including its creator. As Lewis concluded,

> The price of conquest is to treat a thing as mere Nature. Every conquest over Nature increases her domain. The stars do not become Nature till we can weigh and measure them: the soul does not become Nature till we can psychoanalyse her. The wresting of powers *from* Nature is also the surrendering of things *to* Nature. As long as this process stops short of the final stage we may well hold that the gain outweighs the loss. But as soon as we take the final step of reducing our own species to the level of mere Nature, the whole process is stultified, for this time the being who stood to gain and the being who has been sacrificed are one and the same.[16]

But we seem unable to stop short in our drive to make Nature, and ourselves, "natural."[17] The quest, once started, seems self-propagating. It is important that we appreciate the magnitude of this dilemma before proceeding, for it is only in facing the *impossibility* of our situation that the need to search for a resolution becomes apparent. A simple analogy might serve to dramatize the dilemma.

Since our preferences in thought run to problem-solving in all things, let us pose a simple problem: we wish to do something with a troublesome substance so that we can more easily control and use it. Our solution: a container. We can see the principle in action in the simplest of household affairs. Imagine trying to brush one's teeth with paste oozing all across the bathroom counter. Our solution is an obvious one: contain the paste in a tube—an elegant resolution of a messy problem, which gives order to our environment while putting the paste fully at our disposal. But most solutions have innate shortcomings, and this one is no exception. In most households, difficulties arise in use: someone will inevitably squeeze the tube in the middle, trapping some significant portion in the bottom of the tube. With this constriction established, we have two sections in the same tube—a "dualistic" tube, we might say. And if other family members are as willful as the initial constrictor, we may find that the paste can be relocated from day to day, with most lying in the bottom portion on one day, and on the top the next, depending on the preference of the last user.

To relate this to our larger discussion, we might say that the initial "problem" was one of getting a grasp on something which defied easy containment, something large and intractable. We needed a container for the paste, just as our predecessors needed a conceptual container for "everything." Hence, they created "nature." But when the need for a *qualification* of this useful concept emerged, a constriction was established in the nature-tube. Now there could be a section of "everything but," and another for the exception: God or humanity. With that constriction in the conception of "everything," dualism became not only possible, but inevitable, and its eventual refinement to Descartes' "extended matter" and "thinking matter" was prefigured in the initial constriction. But the dilemma we face today arises from our meddling with the constriction. Even though it has been enormously useful to have one section which is deliberately not "like us," which is determinate and manipulable to our purposes, the very success of that device has encouraged us to squeeze more and more of existing substance into the "nature" end. And now we find ourselves slipping through as well, threatened with a final absorption into the mass of nature as the constriction is moved closer and closer to the open end of the tube. Indeed, the dualism appears to be disappearing altogether as the tube becomes once again an undifferentiated, monistic container.

We must bear in mind that the tube was our creation in the first place, our conception and our device. The constriction that appeared was also ours, and the nature-portion of the tube was still a portion of *our* tube: nature was a creature of humanity, a cultural artifact. But we have so escalated the value of that category as the realm of the knowable that it seemed inevitable that we try to squeeze everything into its domain—indeed, to deny that there *is* any other domain: no "supernatural" (God), no "supranatural" (humanity), just Nature.

What was a dualistic device has gradually become, through this manipulation, a monistic one: there is only Nature, and once again, Nature is everything.

Or so it seems. But again, we must remember that nature was itself *our* creation. That we now regard nature as an independent object is because we regard virtually everything as material objects. We seem predisposed to regard whatever we encounter, in direct experience or in thought, as an externally existing entity independent of our own willing and creation. Indeed, objects gain credibility for us as the traces of our authorship are progressively eliminated. But in fact, the absorption of ourselves into Nature is simply the absorption of ourselves into ourselves, or rather, into our own conception of how it "ought" to be. The paradox we encounter, of this perpetual oscillation between the domains of nature and culture, arises from a fundamental error. The dualism cannot actually be resolved, *because it never existed.* The dualism we fret over exists only because of our own decision, not only to constrict the nature-tube into two domains, but to create the container in the first place. One might even say that there is no "nature," and there never has been.

The Rival "Natures"

Nevertheless, we take our concept to be a literal object, and we begin our deliberations from that point. When we detect a disturbance in our protectorate, we rush to discern the problem and proclaim a solution. At the moment, there are different "subspecies" of the concept "nature" in circulation, and some of the conflict identified in the first chapter derives from this confusion.

One view, which is associated with the endeavor to better manage natural environments to protect them from the consequences of industrial society, derives in large part from the understanding of nature that facilitates all scientific questioning. Perhaps a more proximate source of the "resource management" approach is the tradition of scientific agriculture,[18] but the point is simply that all applications of this assumption start with the expectation of a set of objects obeying natural law. The task is to understand how this "system" operates so as to be able to ameliorate any harmful effects. The good intentions of the practitioners need not be called into question. What is at issue here is simply the understanding that they represent, an understanding of "nature-as-object."

Little more need be said about an approach that is so widely accepted other than to repeat that it embodies the understanding that a law-bound nature is the fundamental reality. Indeed, all of our institutional structures assume it. This view is heralded by some as the basic tool of an enlightened "stewardship" of nature, which will permit us to husband all organisms just as we do our domestic animals—not a prospect that everyone would find hopeful, in the age of factory farming. Nevertheless, the expectation is that humankind must use more of the same—that is, direct control through applied science—to help nature (and thus humanity) to survive.

Critics of this approach regard it is inadequate at best, or as the source of the problem at worst. But it is difficult to provide an alternative, and whenever one appears, it is warmly embraced. The fact that Aldo Leopold's famous "land ethic," developed in the 1940s, is still advanced today may indicate the paucity of alternatives. Although there are many variants, the basic position that Leopold advocated—the extension of moral boundaries to include not only other peoples but also other species and even the land itself—seems central. But the idea of extending an ethical system presupposes the existence of creatures that can participate in such a system, and it is here that problems arise. Animals cannot even be consulted about ethical guidelines, much less be expected to appreciate or adhere to them.[19] Proponents, therefore, must devise other means of explaining why animals ought to be included in an ethical system. Many do so by pointing out that sensate beings have ends of their own that cannot be interfered with without harming the creature in question. Hence they may be said to have a "right" to pursue their own ends without interference. Or they may simply have a right to be free from arbitrarily inflicted pain, which makes most of our uses of animals highly questionable.

But others fear that attempting to make animals seem to fit our system is a deceit doomed to be exposed, and attempt instead to suggest that the problem can be circumvented by the realization that *we are them.* That is to say, when we disabuse ourselves of the notion that we are merely skin-encapsulated egos, and realize that we actually have a "field of care"[20] in which we dwell, which makes us literal participants in the existence of all beings, then we will realize that to harm nature is to harm ourselves.[21] Nature is, then, an extended self, and is entitled to the same concern as any other person. This position is what we can call "nature-as-self."[22]

Notice again the lofty intentions of both groups. The first wishes to husband nature efficiently. The second wishes to treat it well, as one would (or should) another person. But despite the variation apparent in the second case, both groups seem to share at least one assumption: that there *is* a thing called nature that needs our help. Both are, in that sense at least, obvious inheritors of the long tradition of nature-thinking. Even though the famous dualism may seem most obvious in the case of those favoring "nature-as-object," it is more accurate to say that both are intrinsically dualistic, and both are simply struggling over where to put the pinch in the nature-tube. The "nature-as-object" group tend to expand the nature-end of the tube, arguing that all can be understood as derivative of natural law and therefore can be amenable to human understanding and control. The "nature-as-self" group tend to squeeze in the other direction, effectively arguing that much of nature is actually "like us" and therefore properly housed in the human or "person" end of the tube. But both believe in the tube. And since that is the case, they are doomed eternally to the "battle of the squeeze."

This battle may be akin to the distinction provided for us by Hans Jonas, between the older "vitalistic" monism and the current "materialistic" monism: is nature more "like us" or are we more "like it?" If the former, then it behooves us to treat it as we would other persons; if the latter, to treat other persons as we do "natural resources." The latter point of view is nearly ubiquitous, but on occasion it has been strongly resisted (are the musings of the sociobiologist really so much more threatening than the "human resources" vocabulary of the bureaucrat?). The nature-as-self position is apt to be dismissed as anthropomorphic, or worse. Yet the burgeoning literature of environmental ethics and rights suggests it is considered acceptable by many.

These two alternatives are not as different as they may seem: if the question that springs to mind in the first instance, of nature-as-object, is "what's in this *for* me," the one that arises in the second is "what is this *to* me," what is my relationship or kinship to this nonhuman nature. In both instances, the centrality of the perceiving human is apparent. Indeed, the self-identification apparent in the nature-as-self position has been described as perhaps the ultimate humanism, since it usurps everything and makes it all "me."[23] However, for our purposes the most interesting facet of this debate is the apparent tension in the underlying dualism: is nature more human, or is the human more natural? A more demanding question might be, "Are these the only alternatives?" Perhaps, if we are questioning the very validity of the dualism that underlies these two contenders, we should we be looking not at the ends of the balance, but at the fulcrum. What would happen, after all, if the center collapsed and the board fell to earth? Would it not be immediately clear that without the fulcrum of dualism, the question of balance cannot even arise?

Expanding the Conversation

In the preceding four chapters we have traced the development of the idea of nature from that of a domain of divine meaning to a separate realm of surfaces stretched taut over a skeleton of reasons. Throughout this transformation, one feature has dominated: an ever-deepening sense of separation between the human subject and the surrounding field of natural objects. Although this separation may be a convenient fiction, it is a difficult one to retain in pure form: it constantly threatens to self-destruct. And yet there seems no way to avoid this conception: it simply seems obvious that there is us, and there is "it," and that the only way we can know about "it" is through the careful explication of the surface details made apparent to our discerning examination. That is only common sense.

What constitutes common sense, however, is a very culturally specific thing. And if, as I suppose, our current environmental dilemma results in part from what seems sensible to us to do to the world, then perhaps we have to admit some non-commonsensical insights into our conversation. In fact, it may be that what we need is a new conversation, one in which the "voices" permitted are not limited to those of practical activity and science.[24] The foregoing overview of the emergence of a new "nature" suggests that we have locked ourselves into a dangerous fiction which, while offering a method of practical power, offers no sense of significance beyond mere exercise of that power. To know "how to" provides no guidance, and no hint of "why to." And, I will suggest, it gives no hint of the core of wild "otherness" in the world that flagrantly belies our claims of exclusivity.

Part 3

The Liberation of Nature

Endangered species are not simply accidents of our way of living. They are the necessary consequences of our way of knowing animals.

CHARLES BERGMAN

7

Nature and the Ultrahuman

But by eroding the old vocabulary, with its rich symbolic overtones, the naturalists had completed their onslaught on the long established notion that nature was responsive to human affairs. This was the most important and most destructive way in which they shattered the assumptions of the past. In place of a natural world redolent with human analogy and symbolic meaning, and sensitive to man's behaviour, they constructed a detached natural scene to be viewed and studied by the observer from the outside, as if by peering through a window, in the secure knowledge that the objects of contemplation inhabited a separate realm, offering no omens or signs, without human meaning or significance. "One of the nicest things about studying wild creatures," thought Gavin Maxwell's zoologist aunt, "is that we are interested in them, while they're not interested in us."[1]

How could wild creatures, constrained by law to attend mechanically to their own affairs, possibly seem interested in us? The presumption of dualism removes even the *possibility* of contemplation in nature, for it effectively removes all *subjects* from nature. If subjectivity, willing, valuation, and meaning are securely lodged in the domain of humanity, the possibility of encountering anything more than material objects in nature is nil.

But while we have successfully banished nature from our immediate domain, we find we must constantly be on guard against the "natural" elements of "human nature," which can still threaten the purity of our private realm. Hence, we make another separation between what have been called the "primary" and "secondary" *human* worlds, so that the merely biological elements of our own domain can at least be kept in check and not be permitted to interfere with our favored cultural ornamentations. Marjorie Grene suggests that "as we acquire our cultural heritage, we come to dwell in it also. We assimilate it to our persons and identify it on the one hand with our primary world and on the other with reality itself. So nature comes to *mean* to us Galilean nature, and the existence of the primary life world is ignored."[2]

Ignored, or even denied. And with that denial, the horizon of our world contracts to a fraction of its original scope. Left beyond, in nature, are the fellow beings with whom we have shared all of our evolutionary existence. And even when contact is apparently made, it is deflected, for as John Berger claims, "animals are always the observed. The fact that they can observe us has lost all significance."[3] They are simply our research project, and while we can certainly observe the "objects" of nature, we can seldom encounter the *other* in nature.

As we have seen, one can respond to this human/nature dualism by attempting to draw the human into the realm of nature, thus effectively eliminating subjectivity altogether; or one can attempt to pull individual species of animals into the realm of the human, and populate our landscape with the pets and puppets that these pseudo-humans inevitably become.[4] But to actually encounter the other beings *as* other, as living subjects of significance, requires some loosening of the conceptual bindings of nature so that subjectivity can flow back in, like water to a scorched garden. This is resisted in the everyday defense of dualism and by the strictures of empirical investigation which dictate that we treat nature "as an invading army treats an occupied country, mixing as little as possible with the inhabitants."[5]

Since it seems only commonsensical that what we are attempting to achieve knowledge of is the objects "out there," one can scarcely imagine how this dualism could be avoided. Yet as we noted earlier, it may only be *because* it is commonsensical that it appears inevitable. And it may be that "common sense" is actually a rarefied faculty, heir to centuries of theoretical explanation. Common sense is an *interpretation* of experience as an encounter of an inner self and outer objects. Yet lived experience does not confirm that interpretation. Erazim Kohak illuminates this through the example of a man who is a smoker searching for an ashtray in the house of a nonsmoker. Obviously his search is doomed from the outset, but he eventually encounters a seashell or a nut dish that serves his purpose. He obviously didn't "find" an ashtray, since there weren't any; he "invented" one. This solution is ignored by common sense, which instead treats the experience as "an encounter with an object out there in the world. Typically, it will report that 'I *found* an ashtray.' But there was no ashtray to be found; the smoker had to constitute it. Here 'I found one' is *an interpretation,* so habitual as to seem 'natural,' but still not a direct report."[6] The ashtray that the smoker claims is "out there" is certainly not merely in his mind, either; it exists *in his experience.* "That is also a crucial point: the world is indeed 'there' *in lived experience,* but that experience is not an ephemeral, transparent nonrealm between a 'subjective' mind and an 'objective' world. Nor is it a passive 'subjective' report of an autonomously existing 'objective' reality. It is *reality,* the only reality that is actually given in experience rather than constructed in speculation."[7]

As should be apparent from the preceding discussion, the entity which we take for granted as an objective reality has, in fact, a complex origin as a social creation. The fact that it *seems* obvious is a function of its absorption into our very expectations of the world, and a function of our willingness to dwell in the world of symbols and abstractions. But when we accept that this "nature" of which we speak is an *interpretation* of our worldly experience, we become open to the question, "what, then, *is* our experience?" What *do* we encounter before we discover "nature?" Surely, some *uncaged* experience of otherness must still be accessible to us?

Maurice Merleau-Ponty suggests that "to return to things themselves is to return to that world which precedes knowledge, of which knowledge always *speaks,* and in relation to which every scientific schematization is an abstract and derivative sign-language, as is geography in relation to the countryside in which we have learnt beforehand what a forest, a prairie, or a river is."[8]

To return to things themselves is to observe them *before* they were "nature,"[9] that is, before they were captured and explained, in which transaction they ceased to be themselves and became instead functionaries in the world of social discourse. Once named and explained, they become *social* creations, and their primordial givenness is subordinated to their social utility.[10]

A forest may be a mythical realm or a stock of unused lumber, but either way, it is able to serve a social function. It is, in that sense, never *itself* but always *ours,* our "system" of distinctions among the worldly phenomena. As noted earlier, it is our habit, and perhaps an inevitable one, to subsequently construe nature as the *source itself.* Yet nature is not the well, but the bucket, and a leaky one at that.

We can certainly know the concept *nature;* as a container, it is ours completely. But the contents can never be known as encountered in experience if we begin with a *denial of* experience. Indeed, we might say that it is through the dismissal of direct, "subjective" experience that we are made vulnerable to the imposition of the social abstraction called nature and the conventions it entails.

But how are we to have any experience of nonobjectified nature if, as social beings, we are inevitably immersed in a world of symbols and abstractions? If we are to address ourselves to the actual experience of others, not to the cultural *explanation* of otherness, we will have to take seriously Merleau-Ponty's adage: "To return to things themselves is to return to that world which precedes knowledge." This statement alludes to the phenomenological method with its act of "bracketing" social conventions so as to achieve as direct an experience of the world as possible. But while this method may seem too arcane (at least as commonly described) for our purposes, it is surely possible for us to take a small step in that direction by asking ourselves what it would mean to actually encounter the world *for the first time.*[11] And perhaps we can best respond to that question by attending to the experience of the person most likely to have such an encounter: the child.[12]

While the idea of a naive infant may be a fiction, it is at least a useful one for our purposes, for it allows us to imagine the encounter of a non-enculturated human with "nature."[13] The point of this exercise is simply to consider what it must be like to encounter something without any conception of what it might be or mean, and perhaps even without the language that would encourage naming. The child, it has been argued, is inclined at first, and indeed for some years, to treat the world as all "me"—not literally and bodily, but in practice: there is a "field of self" which comprises all that is of significance.[14] But there comes a day[15] when the realization occurs that there are things in the world that are *not* me: that there is *other.*

It is difficult for us, accustomed to the self/other distinction, to appreciate the fundamental significance of the discovery. The initial reaction to the awareness of something existing *other than me* is, surely, astonishment: radical astonishment. If, in the narcissistic dawn of a person's awareness, there comes a realization that this "center" is not unique, the world reels. Like Robinson Crusoe finding footprints in the sand, the child awakens to the realization that *there is otherness in the world*—but what *is* it? How is one to make sense of this discovery? And do we ever fully believe that there is something utterly independent of ourselves?

It is obviously impossible for the child, upon first seeing a tree, to say, even to itself, "there's a tree," because there *are* no trees the first time (or words for trees). Trees as a class of beings only exist *after* the repetition of the experience, and the comparison of that experience with those of other people. Trees are a named category of experience, but the first exposure to this phenomenon is inevitably nameless, and is always "one of a kind." It has been said that for animals, every time is the first time. This may be an exaggeration, but it does dramatize the fact that in the absence of symbols that stand for a class of beings, every event is in some sense unique, in contrast to Gaston Bachelard's assertion that the scientist "never sees anything for the first time."[16] A single instance would be meaningless if what is sought is the regularity of the type; any object is one of a set and therefore knowable and predictable. There could never be a knowledge of "dog" unless there were "dogs," for it is the statistical regularities of the group that are indicated by the name. What there is, instead, is a miracle: an animate entity, warm, self-willed, and incontrovertibly *other.* Kathleen Raine speaks of her childhood experience of flowers overhanging her pram, and of her realization that "in the manifold, the innumerable I AM, each flower was its own *I am.*"[17] The child's first such experience might be marked by delight, fear, or amazement. But more important by far is the impact of the realization that there is an other, something in experience which cannot be contained in the self and is, therefore, uncanny—and *wild.* To encounter the wild other, to greet another "I am," is to accept the other's existence in one's life world.

Through the astonishing experience of otherness, the child has some inkling of his or her own being. The shock of encountering something which is other is the shock of being alive: isn't it amazing that there is that, and not just me—and even that there is something rather than nothing.[18] The experience of genuine otherness, the frog in the pond or the bird in the tree, discloses one's own existence. Like the traveler in a foreign land who suddenly becomes aware of his or her own cultural assumptions because they are no longer shared by everyone around, the child becomes aware of itself through the experience of the otherness. Rather than Descartes' "I think, therefore I am," the inarticulate child—who may not "think" at all, in the philosophic sense—discovers that because there is other, because there is "not-me," there is also "me." The other reveals the self.

Self and Other

But the simple realization of the other in nature is far from the end of the story. The self takes form in this initial encounter, but it develops through a subsequent elaboration. The significance of this encounter with the otherness of nature was suggested many years ago by Edith Cobb. Her extensive survey of autobiography led her to speculate about the importance of immersion in nature to the creation of an individual human. Her thoughts have fascinated and bemused several generations of students, but her very independence of mind has made it difficult to know quite what to make of her findings. There seems, however, to be some kinship between her insights and what has been sketched above. For example, she notes that there are frequent references to the "awakening to the existence of some potential, aroused by early experiences of self and world." Further, these "autobiographies repeatedly refer to the cause of this awakening as an acute sensory response to the natural world."[19] She does not suggest that there is any sense of fear or alienation in these reports (and in fairness I should note that she dwells much more on the sense of involvement with the other than I am doing here). For example, one of her authors reports an intense awareness of the material otherness or "Itness" of the nonhuman world, and also a sense of being part of the larger scheme of things in which these others dwell.[20] Overall, though, it is the significance of the natural world, realized in solitude, that is emphasized by all the writers. For only in solitude,[21] in the absence of the demands for social consensus, can the uninterpreted other be encountered and the emergent self escape constraint.

Cobb's evidence points to the importance of an immersion in the uninterpreted world of otherness for the subsequent manifestation of creativity. If a person cannot discover a "world" in the process of self-genesis, he or she will have no choice but to accept that which is provided by others,[22] including the cultural edifice called nature.[23] The opportunity of contrasting the child-self with genuine otherness—not individuality, not "personality," but the full heterogeneity of otherness—provides both the wonder which is the stimulus to the creation of a habitable world and the materials to effect that creation. The implication is that to have the facility to create "worlds" in the cultural medium of words or images, one must first have had the opportunity of creating a world *with the body,* so to speak, first-hand and presemantically. Childhood is the one period in life when one is able to experience a world nonculturally, or at least only semiculturally, before the interpretations of society and the craving to belong to that society foreclose such options. It is a "window of opportunity" that exists before the small human becomes forever a creature of culture and must dwell in the domain of abstractions and representations.

"Experience in childhood is never formal or abstract," Cobb says. "Even the world of nature is not a 'scene,' or even a landscape. Nature for the child is sheer sensory experience, although any child can draw in the wings of his surroundings at will and convert the self into a 'theater of perception' in which he is at once producer, dramatist, and star."[24] Her choice of the term *sensory* was perhaps unfortunate, for it may seem to imply a simple act of sense-reception. But the recognition of otherness is not so much perceptual as it is *sensual.* Perception, as we commonly think of it, is partly an act of comparison, a matching of impressions to established norms so as to "get it right." It is not a passive experience, but a grasping of intended shapes. A sensual experience, however, entails not grasp but abandon, a forgoing of intellect in deference to direct encounter—to become awash in "the inexpressible, the magical effect of pure sensation."[25] Perhaps Gaston Bachelard's comments about poetry are also applicable to nature-experience: the "*density,* which distinguishes the superficial from the profound in poetry, is felt in the transition from *sensory values* to *sensual values.* . . . Only sensual values offer 'direct communication.' Sensory values give only translations."[26]

A sensual experience is always immediate. Carl Jung claimed that "sensuous" pertains to "sensation as a function quite apart from the intellect," and insisted that in this attitude, "the object is neither cognized abstractly nor empathized, but exerts an effect by its very nature and existence."[27] It is, perhaps, a truly radical empiricism, an unadulterated exposure to the otherness of nature. In the case of the sensuous engagement of the child, no objects of reflection are entailed: it is purely a question of the now, the immediate response to a singular, uninterpreted event. This is not to suggest that it is a subconscious event, but rather a preconscious one: consciousness cannot grasp firmly while intentionality is at rest.

Merleau-Ponty stressed the importance of "*being filled with wonder*" at the world, and claimed that "in order to see the world and grasp it as paradoxical, we must break with our familiar acceptance of it and, also, from the fact that from this break we can learn nothing but the unmotivated upsurge of the world."[28] Although he is speaking of the process of world-knowing in general, his comments would be admirably suited to the description of the childhood encounter with the otherness of nature. It involves eliciting a perception of the world "upon which our idea of truth is forever based."[29]

But what does the child find in nature that permits it to make a world? One is tempted to imagine some kind of geometric "building blocks" that could be arranged in a suitable order. Or one might more appropriately think of the child encountering "substances" that can be poured into the receptacles of mind. With a fascination like that shown for water and mud, the child collects fluids of significance toward the eventual creation of the edifice that is the world. A full range of substances must be found before the vagaries of age force the cooling and the congealing which will make that world permanent, the home of that child forever. And each of those substances imbues the final forms with its own essential quality. In the construction of a world the child requires not simply forms in which the wondrous slurry may be poured, but the range of qualities that will give *meaning.* Hence in this process of creation the child needs not just a "formal" imagination, with which we are familiar, but also what Bachelard calls a "material" imagination, which entails "images that stem *directly from matter.* The eye assigns them names, but only the hand truly knows them":[30] a sensual imagery.

Bachelard spent much of his life trying to characterize these images of matter. Indeed, it came as a surprise to many of his scientific colleagues when, in 1938, he published a small volume called *The Psychoanalysis of Fire.* He went on to discuss the imaginative significance of the other classic elements, but it was his treatment of water that evoked valuable reflections from Ivan Illich: "H_2O and water have become opposites," he says. "H_2O is a social creation of modern times, a resource that is scarce and that calls for technical management. It is an observed fluid that has lost the ability to mirror the water of dreams. The city child has no opportunities to come in touch with living water. Water can no more be observed; it can only be imagined, by reflecting on the occasional drop or a humble puddle."[31]

This statement very nearly summarizes the realization thrust upon us in our review of the history of nature. It could be rephrased thus: "Nature and nature have become opposites; Nature is a social creation of modern times, a resource that is scarce and that calls for technical management." We have H_2O, and Nature, but we cannot dream with them, nor can they provide the child with the genesis-experience described by Cobb. Through our conceptual domestication of nature, we extinguish wild otherness even in the imagination. As a consequence, we are effectively alone, and must build our world solely of human artifacts. The more we come to dwell in an explained world, a world of uniformity and regularity, a world without the possibility of miracles, the less we are able to encounter anything but ourselves.

If we can only look to cultural artifacts during our generative process, we must become, in a sense, cybernetic beings, creations of our own technology. Our children learn from "educational" television that nature *is* machinery, and the boundary between flesh and electronics becomes hopelessly blurred. The "otherness" that is required in development will now be of our own making, and therefore not genuinely other at all. The material given the child is preformed by adult society and resists the imagination. Earth, air, fire, and water, the elements which Bachelard considered the "hormones of the imagination," inspire creation. Plastics and stainless steel might be "elements" to a cyborg, but it is improbable they could serve the needs of the organic child's formative imagination. No more can the attempts at idiosyncratic individuality within society satisfy the craving for heterogeneity that can be encountered in the profound otherness of creatures whose manifest abilities prod the imagination toward the brink of wonder.

The Ultrahuman

Perhaps it is difficult for us to appreciate the significance of otherness, since the context of our own existence denies or diminishes genuine heterogeneity. Although we recognize a profusion of objects, our world is essentially homogeneous, just matter: the possibility of some *irregularity* in that plain sheet of being seems contradictory, a breach of the "laws of Nature."[32] That there should be something so far from explanation as to make awe the only appropriate response is almost beyond conception—but perhaps not entirely beyond experience. Indeed, the encounter with radical otherness seems fundamental to certain of our experiences, however rare they have become.

In the face of any phenomenon, we have a choice between explaining it or accepting it. If the former, then we have not seen it, for it becomes just "one of" something else, nothing but another instance of the same old thing. If, on the contrary, it is accepted *in its full individuality,* as a unique and astonishing *event,* our encounter is entirely different, and is perhaps fundamentally religious in the nonecclesiastical sense. In such instances, we experience what Rudolf Otto called the "wholly other": "that which is quite beyond the sphere of the usual, the intelligible, and the familiar, which therefore falls quite outside the limits of the 'canny,' and is contrasted with it, filling the mind with blank wonder and astonishment."[33]

Otto is speaking of the experience fundamental to the religious attitude, not in the sense of dogma and church-going but in the seminal encounter that precedes those. It might be fair to say that the experience of radical otherness is at the base of all astonishment or awe, all "numinous" experience. It is that shock of recognition that generates the acknowledgment of mystery that we can characterize as religious. Otto suggests that "in the last resort it relies on something quite different from anything that can be exhaustively rendered in rational concepts, namely, on the sheer absolute wondrousness that transcends thought, on the *mysterium,* presented in its pure, non-rational form. All the glorious examples from nature speak very plainly in this sense."[34] Significantly, this experience is also reported among the important events of childhood. Richard Coe suggests that "the challenge to the intuition of the child (as to the contemporary metaphysician) is precisely that which *cannot* be assimilated to the human: centipede or stamen or sea anemone," and that "the resistant, unassimilable beauty of *things* is the child's most immediate, most incommunicable experience of the 'sacred.' "[35]

Despite his central preoccupation with religious experience, there seems to be a close kinship between what Otto describes as an encounter with the "wholly other" and the childhood experiences discerned by Cobb and Coe. Moreover, it seems entirely possible to describe the phenomena of which Otto speaks in more secular terms. The English term *wonder* retains some of the force of *miracle,* according to Howard L. Parsons,[36] but even though miracles are said to be "against nature," it is frequently exposure to the realm we call nature that evokes this experience. Not, to be sure, while satisfying our curiosity through explanations of the machinery of nature, but in the sudden exposure to that which goes beyond even the *need* for explanation. R. W. Hepburn suggests that wonder does not imply possession, but rather permits objects to remain "'other' and unmastered."[37] He further claims that "the attitude of wonder is notably and essentially *other-acknowledging.*"[38] In wonder we accept the presence of something entirely distinct and self-possessed. That which evokes wonder is never *ours* in any sense: it is "ultrahuman."*

In his essay "Seeing Nature Whole," John Fowles speaks of our apparent desire to be in control of everything. "Some deep refusal to accept the implications of Voltaire's sarcasm about the wickedness of animals in defending themselves when attacked still haunts the common unconscious; what is not clearly for mankind must be against it. We cannot swallow the sheer indifference, ultrahumanity, of so much of nature."[39] Richard Jefferies, the naturalist whom Fowles credits with the concept of the "ultrahuman,"[40] concluded that because the creatures of nature "were so much to me, I had come to feel that I was as much in return to them. The old, old error: I love the earth, therefore the earth loves me—I am her child—I am Man, the favoured of all creatures. I am the centre, and all for me was made."[41] But the awful truth, according to Jefferies, is that nature does *not* care for us, our plans, our designs, our reasons, or our purposes: it simply *is.* "All nature, the universe as far as we see, is anti- or ultra-human, outside, and has no concern with man."[42] We cannot know it or own it; it is *beyond.*

No doubt this was an extraordinarily difficult conclusion for a devout naturalist, who had treated the world as his cherished community throughout his life. But to realize that despite his devotion to nature, it remains indifferent to him is both devastating and liberating. For it releases nature from being *mine,* a personal interpretation and a wishful thought, to being *its own.* "When at last I had disabused my mind of the

*"Ultra" in this usage refers to that which is "beyond" humanity.

enormous imposture of a design, an object, and an end, a purpose or a system, I began to see dimly how much more grandeur, beauty and hope there is in a divine chaos—not chaos in the sense of disorder or confusion but simply the absence of order—than there is in a universe made by pattern."[43] As we know from Mary Douglas's observations, absence of order is something we tolerate with great difficulty. Without order, there is no dirt, no pollution, no *norms* to point to so as to influence each other's behavior. Yet only when that order—or system, or model called "Nature"—is temporarily abandoned are we able to confront the other *in solido.*[44] The ordering that makes the world seem comprehensible also makes most of it inaccessible.* Without the imposition of one's own *system,*

*There seems to be a curious ambiguity in our treatment of nature with respect to "dirt." On the one hand, it is clear that nature is outside the social order, and is that which has not (at least yet) been given human order: hence, there is culture and there is the great "uncultured," nature. In this sense, nature is dirt, and must be kept outside if the "system" is to be maintained in its purity. We see the consequences of this bit of perversity in every suburb: row upon row of identical housing surrounded by identical lawn, manicured and purified by the ablution rites of chemical lawn maintenance. This spectacle is even defended by laws that punish any who permit "noxious weeds" to grow in their yards. Since these noxious weed laws were originally enacted to protect cattle from consuming dangerous plants, they are clearly no longer needed in suburban neighborhoods. But the fact that they are maintained is evidence that the weeds have become noxious not because of their physiological effects on cattle, but because of their conceptual effect on suburbanites; they are a *pollutant.* They are intrusions into the order of the lawn, and into the domain of human willing. Clearly then, as "natural" (wild) entities which must be excluded, weeds are dirt, as is the rest of nature.

Yet here is the paradox: although we treat nature as the antithesis of order, we also attribute to it a *secret* order. That is, by claiming that there is a reasonable, regular structure behind all the appearances of nature, an order discernible only by the human mind, we also claim it for our own system: we have *ordered* it by claiming privileged access to the "system" within. By "systemizing" nature, we make it ours, a part of the ordered world, a part of culture. So, curiously, we both accept it—the hidden part at least—as an ordered realm, while simultaneously rejecting the "dirty" *manifestations* of that hidden order that are actually encountered in the chaotic domain that strives against the backyard fence. Given this, perhaps the only action a concerned person could take in support of the nonhuman world is to demonstrate a tolerance of the "divine chaos"—including weeds and dirt. To do so would not only expand the habitat of innumerable creatures, but would also confront the system that sustains this organic apartheid.

without the need to assume a *human* organization in the world, what one encounters is simply the other, on its own terms: a "divine chaos." And this, it would seem, is the most difficult thing for us to do. We must make it *ours,* by one means or another. In the most literal sense, we make nature ours in the domestication of plants and animals. In a more personal sense, we make it ours by assuming it is an extended self, or at least a thing *like* ourselves.[45] And in the more general sense, we make it ours by declaring what its form shall be, and by asserting a system of necessity which henceforth shall be known as "Nature."[46] But the divine chaos—or the "green chaos," as John Fowles has it—is exactly the opposite: it is *its own,* and not ours at all. Indeed, "there is a deeper wickedness still in Voltaire's unregenerate animal. It won't be owned, or more precisely, it will not be disanimated, unsouled, by the manner in which we try to own it. When it is owned, it disappears."[47] *Wildness* is the quality of this divine other, and it is wildness that is destroyed *in the very act of "saving" it.* Wildness is not "ours"—indeed, it is the one thing that can *never* be ours. It is self-willed, independent, and indifferent to our dictates and judgments. An entity with the quality of wildness is its own, and no other's. When domestication begins, wildness ends.

Until now, our acts of domestication have entailed the selecting of certain genetic propensities, things that would yield creatures of our choosing. Yet even if genotypes have been selected and manipulated—that is, domesticated—the genes themselves have been beyond our grasp: the genes themselves were wild. With the ability to manipulate DNA, the situation changes. This is, in effect, *the domestication of the gene,* the final assault on the wildness of life. The domestication of the gene exterminates wildness at the source and places all life within the domain of human willing. Nature is domesticated in body, in concept, and finally, one must say, in spirit.

We find ourselves in a dilemma: our actions on behalf of nature seem destined to disappoint, because the *solutions* called forth in response to our despair are animated by the ubiquitous project of global domestication. To "husband" wild animals, or wilderness itself for that matter, is to treat them as domesticates. They are objects to be planned for, matter to be rearranged to maximize its "success" and to ensure it is used "efficiently." And if we—even as "wise stewards"—regard the world as a saintly stockyard, then we have destroyed the very quality we were drawn to save in the first place. Our whole mode of perceiving forces us to domesticate even as we look, and in so doing to deny the possibility of encounter with the other. Every question we ask, every solution we devise, bespeaks mastery, never mystery: they are incompatible. Yet wildness, otherness, *is* mystery incarnate.

It is perhaps not surprising that Thoreau's famous aphorism, "in Wildness is the preservation of the world,"[48] is often misquoted as "in wilderness is the preservation of the world." For wilderness can be regarded as a thing, and as such, susceptible to identification and management. Wildness, however, lies beyond the objects in question, a quality which directly confronts and confounds our designs. At root, it is *wildness* that is at issue: not wilderness, not polar bears, not whooping cranes or Bengal tigers, but that which they as individuals exemplify. These creatures are "made of" wildness, one might say, before they are made of tissue or protein. But perhaps even wildness is an inadequate term, for that essential core of otherness is inevitably nameless, and as such cannot be subsumed within our abstractions or made part of the domain of human willing.[49]

In our everyday assumptions, nature is regarded either as the embodiment of natural laws or as "self"—but either way, it is ours. It cannot be encountered as other, because in a sense it cannot be encountered at all: it is posited as something beyond the human world. The famous dualism conceals the fact that whether we opt for one "side" or the other, we still find *ourselves* and little more. The other, the genuinely *ultra*human being of nature, has become invisible: it has no place.

Yet occasionally, an exceptional adult can still encounter an animal in all its ultrahumanity. In one of her essays, Annie Dillard describes her reaction when, sitting near a pond, she is surprised by a weasel, and it by her. "Our look was as if two lovers, or deadly enemies, met unexpectedly on an overgrown path when each had been thinking of something else: a clearing blow to the gut . . . the world dismantled and tumbled into that black hole of eyes." She did not observe the "behaviors" of the beast, did not retain the proper, adult detachment requisite to the study of nature. Instead, she momentarily lost her self-consciousness and encountered otherness directly and with astonishment. It was, she is certain, no fantasy, no "projection" upon a hapless puppet. "Please do not tell me about 'approach-avoidance conflicts.' I tell you I've been in that weasel's brain for sixty seconds, and he was in mine." What she found was "nothing"—the weasel had no names or ideas to share. So: "What does a weasel think about? He won't say. His journal is tracks in clay, a spray of feathers, mouse blood and bone: uncollected, unconnected, loose-leaf, and blown."[50] The weasel doesn't "think about." There are no abstractions, just a celebration of animate being that can only be fleetingly encountered, and never understood.

Strangeness, according to Owen Barfield, "arouses wonder when we do not understand; aesthetic imagination when we do."[51] Whether our encounter with otherness should lead to one or to the other, or to both, is an open question (unless it should turn out that aesthetic experience is merely domesticated awe). But in any case, neither is possible without that radical otherness which is so often obscured by explanation and neglect. It is unlikely that many of us will ever have the opportunity to duplicate Dillard's encounter, for weasels' eyes have become scarce, and wildness is drained away through "wise management." But if there are not weasels, perhaps there will be something even smaller, less furtive, that a child-sized person could still surrender to. As long as there is hope of contact, some wild spot the gardeners and developers have missed, there is still the chance that some imagination may escape, untamed, into the green chaos.

The French author Alain Robbe-Grillet was especially attuned to the camouflaging tendency of the acculturated consciousness, which renders direct encounter impossible most of the time. "At every instant," he claims, "a continuous fringe of culture . . . is being added to things, disguising their real strangeness, making them more comprehensible, more reassuring." Often the process is complete, but even if the odd piece breaks through the interpretive screen, it can conveniently be dismissed as an absurdity. The task of the artist or writer, then, is to unsettle, to combat this camouflaging process and reassert the observation that "the world is neither significant nor absurd. It *is*, quite simply. That, in any case, is the most remarkable thing about it"[52]—as every child knows, at least once.

It is no accident that we find ourselves, in the end, considering the words of creators such as Dillard who possess the means of evoking sustained images of otherness—and who are uniquely permitted by society to use the essential, metaphoric language. If, as I have suggested, it required the inspired vision of artists of the past to constitute the "things" which occupy the ordered domain of Nature, it will surely require a similar level of inspiration to reconstitute them. The so-called environmental crisis demands not the inventing of solutions, but the re-creation of *the things themselves.*

Of what are things made? Of matter, we are told. But were matter all that existed, there would be no things, just as, without the attention of the smoker, there would be no ashtrays. The "potter's wheel" of objects is the imagination, which Bachelard claimed "is not, as its etymology suggests, the faculty for forming images of reality; it is the faculty for forming images which go beyond reality, which *sing* reality. It is a superhuman faculty."[53] Where are we to find the new things? If the cultural genesis of Nature has provided us with objects which invite a particular kind of activity, and if that activity has proven destructive, then it will not suffice to stare still harder at those same objects. Neither will it be adequate to continue the debate with a vocabulary that admits of no other objects. The language of technological experts cannot accommodate the radical novelty of wildness: indeed, that is just what it has been fashioned to deny. The insinuation of a vocabulary of economy ensures our bondage to the literal and obstructs our access to otherness. Jefferies concluded that "the chief value of books is to give us something to unlearn," for in peeling back the culturally given, one is able to constitute "a sort of floating book in the mind, almost remaking the soul."[54] But that, in a manner of speaking, is also the task of the imaginative writer, except that he or she peels back the ideas written not on paper but on the collective consciousness of the age, thus effecting what Ivan Illich and his colleagues have called an "archaeology of modern certainties."

One of the remarkable features of the human being is its ability to adapt, somctimes at speeds rivaling the mutation rate of viruses. We evolve, so to speak, through metaphor: one day the world is respoken, and a new being is released. Whether or not we have reached this point, whether there actually is the possibility of a re-imagining of things, we cannot know: perhaps our constant mass-media chatter is sufficient to drown out any rival vocabularies. But if it is to come, we must hope that the exotic new variant will be a more benign creature, and one less determined to turn all others into Nature. In wildness is, indeed, the preservation of the world.

Epilogue

First bird watcher: ***What is that?***
Second bird watcher: ***That is only a sparrow.***

A devaluation has occurred. The bird itself has disappeared into the sarcophagus of its sign. The unique living creature is assigned to its class of signs, a second-class mummy in the basement collection of mummy cases.

But a recovery is possible. The signified can be recovered from the ossified signifier, sparrow from **sparrow.**

A sparrow can be recovered under conditions of catastrophe.

WALKER PERCY

When Richard Jefferies concluded, at the end of a life of trying to understand the creatures he so greatly admired, that he could not "know" nature, he liberated himself from a lifelong deceit. In doing so, he also freed nature, as if he were releasing a songbird. He gave up the pretense to knowledge that delimits what a creature may be, and which protects us thereafter from the uncertainties of strangeness: we hide from wildness by making it "natural." Inevitably, what we know is largely our own symbolic representations, which will behave as they were designed to. But of that which they purport to represent, they tell a partial story at best. To accept, at last, his own failure to comprehend the world was no doubt a great disappointment to Jefferies. Yet it was not cause for dismay: "I look at the sunshine and feel that there is no contracted order: there is divine chaos, and, in it, limitless hope and possibilities."[1] Perhaps there is only one conclusion a reflective naturalist can come to: that if we would protect nature from the perils of the "environmental crisis," we must first acknowledge that those perils arose as a consequence of conceptual imprisonment. If we would save the world, we must set it free.

This is a frightening prospect. As members of twentieth-century industrial societies, and as functionaries of technological thought, what we fear most is the loss of control (or at least of the illusion of control). To contemplate actually *letting something be* is very nearly beyond our ability. Perhaps it will remain so. But we must bear in mind that *every* act of control, however well-intentioned, constitutes a continuation and an amplification of the process that has been unfolding since the Renaissance. Leonardo surveyed a wild world, and expunged all potential demons[2] and miracles by declaring them extinct; Francis Bacon later pointed the way to the physical means of universal domestication. It is as if, standing in the midst of lush greenery, we were able to reach down and turn off the water to the entire perimeter, to our surroundings, our "environment." Soon, the wild meadow withered, and the hiding-places of willfulness and caprice were desiccated. In a desert, the whole expanse is ours to survey and order, and mystery is dissipated.

In his study of endangered species, Charles Bergman reflected on the source of our power to control, and on the possibility that this power is immanent in the *kind* of knowledge we seek of nature. Such knowledge, however benignly applied, must inevitably extend our control over nonhuman others.[3] Thus even though our explanations of biological functioning may be employed to sustain the remnants of a fading species in reserves or outright captivity, and thereby provide us the satisfaction of having "saved" one kind of being, the knowledge employed entails the diminishment of that other. Success, therefore, also means failure— successful control over the life and death of the other requires the abrogation of its autonomy. And in legitimizing one mode of knowing over others, we cede to that method the right to define what that creature is and how we shall speak of it.[4] Hence, "when the last of the [California] condors was captured, society suffered a loss—the loss of the wild condor. But something else achieved a victory. The winner was biology. It confirmed its right to define for us what a bird is."[5]

Once defined, the nonhuman other disappears into its new description: it is drawn into a symbolic system which orders and explains, interprets and assigns value. In short, the creature becomes ours as it is made "real" by this assimilation. The wild other disappears the instant it is demystified and saved as a managed resource.

Perhaps every act of salvation has had the same result, but at least some of the earlier rescuers did not deceive themselves of their purpose. Gifford Pinchot, chief proponent of the "conservation" movement in the early part of this century, stated unequivocally that "the first duty of the human race is to control the earth it lives on."[6] Recent disciples have been decidedly circumspect, choosing to cloak their intentions, perhaps even from themselves, with such euphemisms as "wise stewardship" or "sustainable development."[7] But bold platitudes do not assure good results, and to "save," in this context, means little more than to stack canned goods on a pantry shelf, neatly labeled "preserved for future generations." When the other is thus contained, and conceptually pasteurized to remove the taste of wild impurities, there seems no threat of insurrection.

The very language of conservation denies wildness, for that which cannot be contained by measurement has no place in that vocabulary.[8] And although we may attempt to imprison it in "wilderness areas" or "game farms," wildness, as Thoreau knew, cannot be forever repressed:

> I love to see the domestic animals reassert their native rights,—any evidence that they have not wholly lost their original wild habits and vigor; as when my neighbor's cow breaks out of her pasture early in the spring and boldly swims the river, a cold, gray tide, twenty-five or thirty rods wide, swollen by the melted snow. It is the buffalo crossing the Mississippi. This exploit confers some dignity on the herd in my eyes,—already dignified. The seeds of instinct are preserved under the thick hides of cattle and horses, like seeds in the bowels of the earth, an indefinite period.[9]

That reassertion of "native rights" becomes increasingly difficult with the rise of factory farming, the expansion of scientific agriculture in the guise of wildlife management, and the impending domestication of the gene. But perhaps acknowledgment of the quality of resistance that Thoreau called "wildness" and Jefferies called "divine chaos" is prerequisite to any serious defense of life on Earth.

One thing at least becomes apparent from these reflections: that "Nature" is a perilous device, all too easily employed in the quest to dominate others. C. S. Lewis warned that "we reduce things to mere Nature *in order that* we may 'conquer' them. We are always conquering Nature, because 'Nature' is the name for what we have, to some extent, conquered."[10] To consign something to Nature—including ourselves—is to submit it to domination and control. Yet in a sense, Nature is also a mode of concealment, a cloak of abstractions which obscures that discomforting wildness that defies our paranoid urge to delineate the boundaries of Being. That which will not be named cannot be controlled. Yet even a rigorous nomenclature may not constitute a permanent constraint, for as Walker Percy suggests, "a sparrow can be recovered under conditions of catastrophe."[11]

What catastrophe might set the sparrow free of the "ossified signifier" that seals it in Nature? Perhaps if, as Richard Rorty suggests, cultural change is a matter of changing vocabularies rather than unearthing truths, the required catastrophe must entail a challenge to our habitual language, and thus to common sense. "Nature," he suggests, "is whatever is so routine and familiar and manageable that we trust our own language implicitly."[12] It is the *unfamiliar* that shakes that complacency and makes us doubt the adequacy of conventional vocabularies.[13]

The unfamiliar can be encountered, on occasion, among nonhuman others, if they are permitted to confront us in all their startling strangeness. It is encountered also in language, most dramatically when a new metaphor intrudes upon the safe domain of literal, commonsensical things. It is, as Rorty says, as if one paused in mid-sentence to make a face, or to slap the listener; it doesn't make *sense*. That is its strength, for such small catastrophes break the "crust of convention"[14] sufficiently for doubt to enter, possibilities to develop, and the occasional sparrow to escape.

If wild otherness is to be sustained, those who have encountered it must not forfeit their right of description to the conventions of domesticating speech. And if, as Rorty suggests, "a talent for speaking differently, rather than for arguing well, is the chief instrument of cultural change," then perhaps we need to pay greater respect to those who, like Charles Bergman, John Fowles, Annie Dillard, and Barry Lopez, clearly possess that gift in the measure necessary to retrieve the "green chaos" from Nature. It is speakers such as these that may help us acquire the vocabulary needed to accommodate wildness and extinguish the technological flashfire of planetary domestication.

Notes

Preface

1. Stephen Fox's history of the American conservation movement, *John Muir and His Legacy: The American Conservation Movement* (Boston: Little, Brown, 1981), provides an excellent overview.

2. See, e.g., John Livingston, *The Fallacy of Wildlife Conservation* (Toronto: McClelland & Stewart, 1981); David Ehrenfeld, *The Arrogance of Humanism* (New York: Oxford University Press, 1978); Carolyn Merchant, *The Death of Nature* (New York: Harper & Row, 1980); Morris Berman, *The Re-enchantment of the World* (Ithaca: Cornell University Press, 1981); Paul Shepard, *Nature and Madness* (San Francisco: Sierra Club Books, 1982); Max Oelschlaeger, *The Idea of Wilderness* (New Haven: Yale University Press, 1991).

3. This is, of course, a topic that has been explored by others, although not, I think, in quite the manner in which I approach it. See, e.g., Berman, *The Re-enchantment of the World;* R. G. Collingwood, *The Idea of Nature* (London: Oxford University Press, 1960); William Leiss, *The Domination of Nature* (Boston: Beacon Press, 1974); Merchant, *The Death of Nature.*

Chapter 1: The Social Use of Nature

1. See, e.g., Fairfield Osborne, *Our Plundered Planet* (Boston: Little, Brown, 1948); Rachel Carson, *Silent Spring* (Boston: Houghton Mifflin, 1962); Clarence Glacken, *Traces on the Rhodian Shore* (Berkeley: University of California Press, 1967); Oelschlaeger, *The Idea of Wilderness*.

2. For the moment, I will accept the common usage of *environment* or *natural environment* as a near-equivalent of *nature.* However, I do so with considerable misgivings, since *environment* is a highly anthropocentric term that encourages the notion that nature is strictly "ours"—i.e., "our environment." Furthermore, since *environment* literally means *surroundings,* it is obviously a rather bland and inadequate term to use in reference to nature.

3. *Oxford English Dictionary,* 2d ed. (Oxford: Clarendon Press, 1989), 12:43.

4. See Mary Douglas, *Purity and Danger* (London: Routledge & Kegan Paul, 1966), p. 35:

> Where there is dirt, there is system. Dirt is the by-product of a systematic ordering and classification of matter, in so far as ordering involves rejecting inappropriate elements. This idea of dirt takes us straight into the field of symbolism and promises a link-up with more obviously symbolic systems of purity.

For a discussion tailored more directly to the question of natural environments, see her essay "Environments at Risk," in *Ecology: The Shaping Enquiry,* ed. Jonathan Benthall (London: Longman, 1972), pp. 129–45.

5. Douglas, *Purity and Danger,* p. 3.

6. Ibid., p. 113.

7. Stephen Cotgrove, in *Catastrophe or Cornucopia* (Chichester: John Wiley & Sons, 1982), p. 33, concludes:

> What the environmentalists are saying is that the society we have got is bad: that the way we behave is against "Nature"; our children will suffer, and time is running out. They want a more convivial society in which community replaces unbridled individualism, in which people are valued as individuals and not for what they have achieved, in which work is humanly satisfying, and in which individuals have more say in the decisions which affect their lives. . . . [The industrialists], by contrast, believe in a society dedicated to the production of wealth, in which individuals are set free from constraints and restrictions, and are rewarded for what they achieve; in which efficiency and the needs of industry are the touchstones of policy.

8. Daniel Simberloff, "A Succession of Paradigms in Ecology: Essentialism to Materialism and Probabalism," *Synthese* 43 (1980): 30.

Recent interest in "chaos theory" has prompted still other critiques of the idea of balance and stability. See, e.g., Daniel B. Botkin, *Discordant Harmonies: A New Ecology for the Twenty-First Century* (New York: Oxford University Press, 1990).

9. Don E. Marietta, Jr., "The Interrelationship of Ecological Science and Environmental Ethics," *Environmental Ethics* 1 (1979): 197. Marietta outlines an interesting phenomenological approach to the problem, but is criticized by Tom Regan ("On the Connection between Environmental Science and Environmental Ethics," *Environmental Ethics* 2 [1980]: 363–66) for running afoul of the "is/ought" impasse. For a review of the philosophical approach to this kind of argument from ecology, see J. Baird Callicott, "Hume's Is/Ought Dichotomy and the Relation of Ecology to Leopold's Land Ethic," *Environmental Ethics* 4 (1982): 163–74.

10. Marietta, "Interrelationship," p. 201.

11. Robert Henry Peters, "From Natural History to Ecology," *Perspectives in Biology and Medicine* 21 (1980): 202. It appears that even recent challenges to conventional ideas of how nature works, such as Botkin's *Discordant Harmonies* (see p. 192), do not require any relinquishing of the quest for global control of nature.

12. The example given here is a reworked version of my essay entitled "Constructing the Natural: The Darker Side of the Environmental Movement," originally published in *North American Review* 270, 1 (March 1985): 15–19, and reprinted here by permission.

13. Most of the details in this section are derived from C. S. Holling, "Resilience and Stability of Ecological Systems," *Annual Review of Ecology and Systematics* 4 (1973): 1–23. It would be unfair, however, to associate him in any way with the speculative discussion in which I use his examples. For further consideration of the spruce budworm issue, see C. D. Fowle, A. P. Grima, and R. E. Muan, eds., *Information Needs for Risk Management* (Toronto: University of Toronto Institute for Environmental Studies Monograph 8, 1988), pp. 157–75.

14. For a discussion of such possibilities, see, e.g., Lewis Thomas, *Lives of a Cell* (New York: Bantam, 1974), and D. Reanny, "Extrachromasomal Elements as Possible Agents of Adaptation and Development," *Bacteriological Review* 40 (1976): 552–90.

15. Marietta, "Interrelationship," p. 205.

16. Marshall Sahlins, *The Use and Abuse of Biology: An Anthropological Critique of Sociobiology* (London: Tavistock, 1977), pp. 72–73. Sahlins sums up the dilemma in his observation that "we seem unable to escape from this perpetual movement, back and forth between the culturalization of nature and the naturalization of culture. It frustrates our understanding at once of society and of the organic world" (p. 105).

Chapter 2: Nature and Norm

1. I will be relying mainly on C. S. Lewis's treatment of *nature* in his *Studies in Words,* 2d ed. (Cambridge: Cambridge University Press, 1967), which is one of the more accessible accounts of how the term has been defined and used. His *The Discarded Image* (Cambridge: Cambridge University Press, 1967) is also very helpful in its discussion of the medieval "model" of nature. (See the bibliographical essay at the end of the book for other titles that can provide the basis of a broader exploration of the concept *nature.*) This is not to suggest that Lewis's is the only treatment available: Collingwood's *The Idea of Nature* is an obvious alternative, but one that entails a more formally philosophic approach than seems appropriate here, as does Richard Rorty's highly regarded book, *Philosophy and the Mirror of Nature* (Princeton: Princeton University Press, 1979).

2. Lewis, *Studies in Words,* p. 61.

3. Ibid., pp. 61–62.

4. Ibid., p. 37.

5. Lewis, *The Discarded Image,* p. 37.

6. Lewis, *Studies in Words,* p. 43.

7. Ibid., p. 46.

8. For a good introduction to the analysis of advertising, see William Leiss, Stephen Kline, and Sut Jhally, *Social Communication in Advertising* (Toronto: Methuen, 1986). Section 3 contains a concise introduction to the use of semiological analysis (pp. 150–69) and some comment on the use of images of nature (e.g., pp. 190, 194).

9. In introducing the work of Roland Barthes here, it is not my intention to discuss his ideas on semiotics, nor to take any side in the debate over its merits. Suffice it to say that he is a significant figure who has many admirers and many critics, and some who are both (Norman Bryson, whose work is discussed in Chapter 4 below, is a case in point). However, Barthes is useful in our discussion because he nicely exemplifies the concern that the domains of humanity (freedom) and nature (necessity) be kept separate, and clearly articulates the dangers he perceives in allowing any mixing of the two. For a more thorough introduction to Barthes's work, see his *Elements of Semiology* (New York: Hill & Wang, 1968).

10. This example is drawn from Barthes's *Mythologies* (London: Paladin, 1973), pp. 121–24, but the reader should refer to his text for a proper explanation of the genesis of sign and myth. While much of what Barthes says is important to a consideration of the social construction of reality, what is significant for our purposes is his contention that, in this process, social concepts are secreted in nature, resulting in a new or pseudo-nature that is, in fact, fully historical.

11. This observation is also stressed by Steven Shapin and Simon

Schaffer in their study of the emergence of English empiricism, *Leviathan and the Air-Pump: Hobbes, Boyle, and the Experimental Life* (Princeton: Princeton University Press, 1985). See, e.g., p. 23:

> There is nothing so given as a matter of fact. In common speech, as in the philosophy of science, the solidity and permanence of matters of fact reside in the absence of human agency in their coming to be. Human agents make theories and interpretations, and human agents may therefore unmake them. But matters of fact are regarded as the very "mirror of nature." Like Stendhal's ideal novel, matters of fact are held to be the passive result of holding a mirror up to reality. What men make they may unmake; but what nature makes no man may dispute. To identify the role of human agency in the making of an item of knowledge is to identify the possibility of it being otherwise. To shift the agency onto natural reality is to stipulate the grounds for universal and irrevocable assent.

12. Barthes, *Mythologies*, p. 108.

13. Ibid., pp. 154–55. See also Norman Bryson's criticism of this interpretation in *Vision and Painting: The Logic of the Gaze* (New Haven: Yale University Press, 1983), pp. 72–77.

14. Peter Berger and Thomas Luckmann, *The Social Construction of Reality* (Garden City: Anchor Books, 1967), p. 89. In this instance, the authors are speaking of the process of reification, in which an institutional order is objectified by making it seem to have a nonhuman origin. Although their modes of explanation differ, the point being made by both Barthes and these authors seems essentially the same.

15. Arthur O. Lovejoy and George Boas, *Primitivism and Related Ideas in Antiquity* (New York: Octagon Books, 1973), p. 111.

16. Ibid., p. 109.

17. For a discussion of this with respect to environmentalism, see Douglas Torgerson, "The Paradox of Environmental Ethics," *Alternatives* 12 (Winter 1985): 26–36.

18. Lovejoy and Boas, *Primitivism*, p. 110.

19. Barthes, *Mythologies*, p. 108.

20. While Barthes's choice of words makes him especially helpful for our purposes, his observation is by no means unique. Indeed, the seventeenth-century intellectual Blaise Pascal may have had much the same thing in mind when he wrote, "Custom determines all justice, for this simple reason that it is accepted; that is the mystic foundation of its authority. Whoever will trace justice back to its principles destroys it" (Blaise Pascal, *Selections from The Thoughts*, Arthur H. Beattie, trans. & ed. [Arlington Heights, Ill.: Harlan Davidson, 1965], p. 55). When something is received as simply a *given*, it appears inevitable. But if it is challenged and its authority exposed (as by Barthes's "progressive humanist"), its power is destroyed.

21. Walker Percy, *The Thanatos Syndrome* (New York: Ivy Books, 1987), p. 80.

22. Karl Löwith, *Nature, History, and Existentialism and Other Essays in the Philosophy of History* (Evanston: Northwestern University Press, 1966), p. 20. Löwith goes on to suggest that "the earth has not become more familiar to us since we have become capable of covering immense distances in a short time. The more we plan globally and exploit the earth technically, the further nature recedes in spite of all our technical seizure of it" (pp. 20–21). It is, perhaps, a similar insight that prompted C. S. Lewis's conclusion that "we reduce things to mere Nature *in order that* we may 'conquer' them. We are always conquering Nature, because 'Nature' is the name for what we have, to some extent, conquered" (*The Abolition of Man* [New York: Macmillan, 1955], pp. 82–83).

23. Löwith, *Nature*, p. 21.

24. Barthes, *Mythologies*, p. 108.

25. *Concise Oxford Dictionary*, 4th ed. (Oxford: Oxford University Press, 1960), p. 580.

26. Paul Edwards, ed., *Encyclopedia of Philosophy* (New York: Macmillan, 1967), 3:69–70.

27. Charles Hartshorne, *Beyond Humanism: Essays in the Philosophy of Nature* (Gloucester: Peter Smith, 1975), pp. 1–2.

28. A much more explicit connection between humanism and the environmental crisis was made by David Ehrenfeld in *The Arrogance of Humanism*.

29. I should point out that this term, which refers to one particular approach to the defense of animals, is now used by journalists as a derogatory category for virtually anyone who demonstrates sympathy for nonhuman animals. However, once a term has been repeated often enough, it becomes almost impossible to avoid, and I do so here—with serious reservations—simply because I assume its colloquial meaning will be readily apparent, and that it will suggest to the reader the entire range of discussion over human-animal relationships.

30. Robert Fulford, "Monkey Business," *Saturday Night* 101 (January 1986): 7. I should note that I do not single out Mr. Fulford as an extreme example of this kind of argument, but rather as one of the more thoughtful and articulate. Because he writes so well, he provides a very clear description of a position that is widely held.

31. James Serpell, *In the Company of Animals: A Study of Human-Animal Relationships* (Oxford: Basil Blackwell, 1986). See esp. chap. 2.

32. Christopher Lasch's searching exploration of the assumptions of contemporary American society led him to conclude that "the capacity for loyalty is stretched too thin when it tries to attach itself to the hypothetical solidarity of the whole human race. It needs to attach itself to specific people and places, not to an abstract ideal of

universal human rights" (*The True and Only Heaven: Progress and Its Critics* [New York: Norton, 1991], p. 36).

33. Interestingly, Fulford suggests that the sole positive outcome of the "animal rights" discussion might be the realization that "at the heart of the movement is a challenge to humankind to define itself; by blurring the distinction between non-humans and humans, the movement forces us to think again about what we are" ("Monkey Business," p. 7). The question of definitions is certainly central, although it is somewhat ironic that the very idea of animal "rights"—an idea that seems to presuppose a humanist interpretation of justice and individuality—amounts to a kind of subterfuge to draw certain nonhumans into the protected circle of human self-definition. In the broader discussion of "animal liberation" (a term coined by philosopher Peter Singer in his book of that title), the notion of rights need not arise, although some notion of justice often does.

34. Lewis, *Discarded Image*, p. 42.

Chapter 3: The Purification of Nature

1. Giovanni Pico Della Mirandola, *Oration on the Dignity of Man* (Los Angeles: Gateway Editions, 1956), p. 3.

2. On the question of earlier points of change, see, for example, the discussion of the so-called Twelfth-Century Renaissance provided by M.-D. Chenu, *Nature, Man, and Society in the Twelfth Century* (Chicago: University of Chicago Press, 1968).

3. Ernst Cassirer, *The Individual and the Cosmos in Renaissance Philosophy* (Philadelphia: University of Pennsylvania Press, 1972). For another treatment of the centrality of the Renaissance in environmental attitudes, see John Opie, "Renaissance Origins of the Environmental Crisis," *Environmental Review* 11 (Spring 1987): 3–17.

4. Empathy is defined as the "power of identifying oneself mentally with (and so fully comprehending) person or object of contemplation" (*The Oxford Dictionary of Current English* [Oxford: Oxford University Press, 1984], p. 240).

5. Cassirer, *Individual and Cosmos*, p. 148.

6. John Herman Randall, Jr., *The Making of the Modern Mind* (Boston: Houghton Mifflin, 1940), p. 28.

7. See Chenu, *Nature, Man, and Society*, p. 14.

8. Cassirer cites the example of Campanella, a Renaissance figure who sought a new metaphysic through the analysis of signs: "For him, 'to know' means simply to read the divine signs that God has written into nature." Yet while this is the goal, the means remains empathetic:

> The bond that holds together the innermost recesses of nature and that joins nature to man is still conceived of completely as a magical-mystical bond. Man can only understand nature by inserting

his own *life* into it. The limits of his feeling for life, the barriers to a directly *sympathetic feeling* of nature, are at the same time the limits of his *knowledge* of nature.

This is in direct contrast with the path of interpretation taken by Cusanus and later by Leonardo, Galileo, and Kepler, a path that "is not satisfied with the imagistic and sensible force of the signs in which we read the spiritual structure of the universe; instead, it requires of these signs that they form a system, a thoroughly ordered whole" (*Individual and Cosmos,* p. 54).

9. Chenu, *Nature, Man, and Society,* p. 102. Of the medievals' approach to knowing nature, Michel Foucault observes that "there are no resemblances without signatures. The world of similarity can only be a world of signs." And of course, one can only learn of these underlying similarities by studying the signatures: "To search for a meaning is to bring to light a resemblance. To search for the law governing signs is to discover the things that are alike" (*The Order of Things* [New York: Vintage Books, 1973], pp. 26, 29). Foucault's work is of course very useful to the study of the transformation in the concept of nature, but those who have dipped toes into his arcane waters will understand why I am reluctant to pursue this topic beyond the fairly minimal level that seems useful for our discussion.

10. Stephen Toulmin and June Goodfield, *The Architecture of Matter* (Harmondsworth: Penguin Books, 1965), p. 91.

11. Alan of Lille and Hughe of Saint-Victor, respectively, cited by Chenu, *Nature, Man, and Society,* p. 117.

12. For a discussion of the medieval outlook as it relates to human attitudes to the natural environment, see Berman, *The Re-enchantment of the World;* also see Theodore Roszak, *Where the Wasteland Ends* (Garden City: Doubleday, 1972), for further discussion of the contrast between the older conception of nature with the newer, post-seventeenth-century view.

13. Chenu, *Nature, Man, and Society,* p. 103.

14. Ibid., p. 123.

15. Ibid., pp. 123–24.

16. Randall, *The Making of the Modern Mind,* pp. 98–99.

17. For an interesting and concise overview of this topic, see Umberto Eco, *Art and Beauty in the Middle Ages* (New Haven: Yale University Press, 1986). See esp. chap. 5, "Symbol and Allegory."

18. For further explanation, see David Knowles, *The Evolution of Medieval Thought* (New York: Vintage Books, 1962), p. 55.

19. "Augustine's 'sign' belonged on the level of his psychology of knowledge," and consequently it was "the knower himself who was the principle and rule of the 'sign'; it was he who gave the 'sign' its value, over and beyond any objective basis in the nature of things" (Chenu, *Nature, Man, and Society,* p. 125).

20. This is also the tendency that C. S. Lewis refers to as "the great

process of Internalisation" in which "century by century, item after item is transferred from the object's side of the account to the subject's." See *The Discarded Image*, p. 215.

21. Chenu, *Nature, Man, and Society*, p. 126.

22. Ibid., p. 131 (emphasis added).

23. Ibid., p. 127.

24. On this, see Lowith, *Nature, History, and Existentialism*, p. 29:

> With the dissolution of these two ancient convictions—the classical and the Christian—historicism and existentialism came into being. If the universe is neither eternal and divine (Aristotle), nor contingent but created (Augustine), if man has no definite place in the hierarchy of an eternal or created cosmos, then, and only then, does man begin to "exist," ecstatically and historically.

25. William James, cited by Carl Jung, *Psychological Types* (Princeton: Princeton University Press, 1976), p. 300.

26. In *Psychological Types*, pp. 410–11, Jung says that when taking an abstract attitude to an object,

> my interest does not flow into the whole, but draws back from it, pulling the abstracted portion into myself, into my conceptual world, which is already prepared or constellated for the purpose of abstracting a part of the object. (It is only because of a subjective constellation of concepts that I am able to abstract from the object.)

27. Ibid., p. 292.

28. Ibid., p. 293.

29. *The Oxford Dictionary of English Etymology* (Oxford: Oxford University Press, 1966), p. 5.

30. *Oxford English Dictionary*, 1:56. See also Jung's definition in *Psychological Types*, pp. 409–10:

> Abstraction . . . is the drawing out or singling out of a content (a meaning, a general characteristic, etc.) from a context made up of other elements whose combination into a whole is something unique or individual and therefore cannot be compared with anything else. . . . Abstraction, therefore, is a form of mental activity that frees this content from its association with the irrelevant elements by distinguishing it from them, or in other words *differentiating* it.

31. C. S. Lewis wished "to persuade the reader not only that this Model of the Universe is a supreme medieval work of art but that it is in a sense the central work, that in which most particular works were embedded, to which they constantly referred, from which they drew a great deal of their strength" (*Discarded Image*, p. 12).

32. Jung, *Psychological Types*, p. 410.

33. It should be noted that the parallel drawn here between the introverted "abstracter" and the extraverted "empathizer" is not entirely consistent with Jung's representation. Indeed, one could well

imagine that the "projection" of qualities would seem less distressing to the introvert than to the extravert, since the latter is said by Jung (p. 310) to be generally contemptuous of "secondary qualities" and insistent on a reduction to purely empirical observations. However, the use to which Jung's insights are employed in our discussion centers on the contrast between the *empathizing tendency* of the extravert and the *abstracting* one of the introvert, a contrast that seems analogous to the rejection of the symbol-laden nature of the medieval period in favor of the internalized and rational Nature of the Renaissance and later. There is, however, no inference that the parallel is exact in all details; it is an analogy drawn for the sake of clarification, and one that permits certain connections to be drawn later in the text.

34. Hans Jonas, *The Phenomenon of Life: Toward a Philosophical Biology* (Chicago: University of Chicago Press, 1982), p. 35 (emphasis added).

35. For an elaboration of this, see E. A. Burtt, *The Metaphysical Foundations of Modern Science,* 2d ed. (Atlantic Highlands: Humanities Press, 1980), p. 100.

36. Ibid., p. 89.

37. Jonas, *The Phenomenon of Life,* pp. 34–35.

38. See J. H. van den Berg, *The Changing Nature of Man* (New York: Delta, 1975), pp. 210ff; see also van den Berg, *A Different Existence* (Pittsburgh: Duquesne University Press, 1972), pp. 18–20, p. 64.

39. Mary Douglas, *Purity and Danger* (London: Routledge & Kegan Paul, 1966), p. 40.

40. S. Drake, ed. and trans., *Discoveries and Opinions of Galileo* (Garden City: Doubleday, 1957), p. 274.

41. Cassirer, *Individual and Cosmos,* pp. 54–55.

42. For further discussion of this event, see Neil Evernden, *The Natural Alien: Humankind and Environment* (Toronto: University of Toronto Press, 1985).

43. Douglas, *Purity and Danger,* p. 2.

44. See Jonas, *Phenomenon of Life,* p. 10: "This denuded substratum of all reality could only be arrived at through a progressive expurgation of vital features from the physical record and through strict abstention from projecting into its image our own felt aliveness."

45. Ibid., p. 9.

46. *Webster's Ninth New Collegiate Dictionary* (Springfield: Merriam-Webster, 1983), p. 90.

47. This tendency has been effectively revealed by feminist writers such as Donna Harraway, *Primate Visions: Gender, Race, and Nature in the World of Modern Science* (New York: Routledge, 1987). See also Linda Fedigan, *Primate Paradigms: Sex Roles and Social Bonds* (Montreal: Eden Press, 1982).

48. Science-derived terms such as *fetus* which have only recently found their way into the vocabulary of self-description are poignant indicators of the power of "Nature-terms" in the redefinition of human affairs. Barbara Duden's studies of "history beneath the skin" are particularly revealing. See, e.g., her book, *The Woman beneath the Skin: A Doctor's Patients in Eighteenth-Century Germany* (Cambridge, Mass.: Harvard University Press, 1991).

49. Burtt, *Metaphysical Foundations,* pp. 238–39.

Chapter 4: From nature *to* Nature

1. Ernst Cassirer speaks of "Leonardo's influence on Galileo," and specifically of the formulation of individual laws of nature, on which point "Galileo bases himself directly on Leonardo" (*Individual and Cosmos,* p. 156).

2. Ibid., p. 152.

3. Ibid., p. 51.

4. Ibid, pp. 57, 58 (emphasis added). For Leonardo's own words on this, see Pamela Taylor, ed., *The Notebooks of Leonardo Da Vinci* (New York: Mentor Books, 1960), pp. 88, 196–97.

5. Cassirer, *Individual and Cosmos,* p. 156.

6. Ibid., p. 164.

7. For a full discussion of the Renaissance transformation in art and its significance, see David Summers, *The Judgement of Sense* (Cambridge: Cambridge University Press, 1987).

8. Löwith, *Nature, History, and Existentialism,* p. 21.

9. Cassirer, *Individual and Cosmos,* p. 153.

10. In *The Crisis of European Science and Transcendental Phenomenology* (Evanston: Northwestern University Press, 1970), pp. 48–50, Edmund Husserl writes:

> Thus it could appear that geometry, with its own immediately evident a priori "intuition" and the thinking which operates with it, produces a self-sufficient, absolute truth which, as such—"obviously"—could be applied without further ado. That this obviousness was an illusion . . . that even the meaning of the application of geometry has complicated sources: this remained hidden for Galileo and the ensuing period. Immediately with Galileo, then, begins the surreptitious substitution of idealized nature for prescientifically intuited nature.

Interestingly, Galileo's "forced fit" of perfect geometric shapes on an irregular nature is still being tailored; the new computer-driven quest for "fractals" seems set to reassert the reality of geometric shapes in nature. For an introduction to this topic, see Benoit B. Mandelbrot, *Fractals: Form, Chance, and Dimension* (San Francisco: W. H. Freeman, 1977).

11. Husserl, *Crisis,* p. 51.

12. Martin Heidegger, *Basic Writings* (New York: Harper & Row, 1977), pp. 250, 251, 252.

13. Cassirer, *Individual and Cosmos,* p. 159.

14. Ibid., p. 58.

15. Ibid., p. 158.

16. Van den Berg, *A Different Existence,* p. 40.

17. On the question of the "true copy" of a presumably singular reality, see Norman Bryson's *Vision and Painting,* particularly chap. 2.

18. C. S. Lewis observes that "Nature, for Chaucer, is all foreground; we never get a landscape" (*Discarded Image,* p. 101). One can scarcely imagine that such a person would have great need for linear perspective—or that he would see nature as we do.

19. Rainer Maria Rilke, *Selected Prose,* trans. by G. Craig Houston (London: Hogarth Press, 1954), 1:3–4. I do not mean to imply that the *Mona Lisa* was itself the principal instrument in Leonardo's redefinition of nature. Indeed, the work would not have been available to his Italian counterparts, since it was completed over many years, many of them while he was living in France where, upon his death, it became the property of the French monarchy. Rather, this painting is singled out because, better than others of the time, it epitomizes the transformation of reality into an "outer" world, a geometric prison for rule-bound functionaries, and an "inner" world of the purpose-giving human individual.

20. Ibid., p. 4.

21. Loneliness is a subordinate but omnipresent theme in the evolution of human consciousness. When one imagines what it must have been like to live in a world in which the divinities are immanent in the surrounding world, the contrast with our own existence is startling. It is not an exaggeration to say that for ancient peoples, one was never alone. In virtually every entity, there was the possibility of sentience. It was truly a living world. But as the powers of Being were gradually withdrawn from earth and invested in a singular and extraterrestrial divinity, that companionship diminished. Although it was still possible to feel the presence of God at all times, it was a transcendental God, not one immanent in the world. And with Nietzsche's final revelation of the death of God—and His succession by Humanity to the throne of omniscience—it became our fate to be utterly alone. Loneliness is the price we pay for the confiscation of power. And loneliness, according to van den Berg, "is the nucleus of psychiatry" (*A Different Existence,* p. 105).

22. Rilke, *Selected Prose,* 1:5.

23. Cassirer, *Individual and Cosmos,* p. 162.

24. Ibid., p. 67. Cassirer suggests that Galileo perceives a very similar role for the scientist:

The whole science of nature, according to Galileo, rests upon this new relationship between understanding *(discorso)* and sense, between experience and thought. And it is a relationship which, very clearly, is exactly analogous to the relation that exists in the Renaissance theory of art between the imagination of the painter and the "objective" reality of things. (p. 164)

25. Ibid., p. 158.

26. Leonardo, cited in E. H. Gombrich, *Norm and Form,* 2d ed. (London: Phaidon, 1971), p. 112 (emphasis added).

27. Ibid., pp. 116, 117.

28. Anthony Storr, *Solitude: A Return to the Self* (New York: Free Press, 1988), pp. 88–89. In the section alluded to here, Storr is principally reviewing Carl Jung's discussion of the association of introversion with the tendency to abstraction and extraversion with the tendency to empathy. (See Jung, *Psychological Types,* pp. 289–99. See also the discussion of Jung's work in Chapter 3 above.)

29. Storr, *Solitude,* p. 89.

30. Cassirer, *Individual and Cosmos,* p. 170.

31. Wilhelm Worringer, *Abstraction and Empathy* (Cleveland: World Publishing, 1967), pp. 133–34.

32. Ibid., pp. 16–17.

33. Cassirer, *Individual and Cosmos,* p. 155.

34. Samuel Y. Edgerton, Jr., *The Renaissance Rediscovery of Linear Perspective* (New York: Basic Books, 1975), pp. 7, 24.

35. Ibid., p. 30.

36. Johannes Kepler, cited by Arthur Koestler, *The Sleepwalkers* (Harmondsworth: Penguin Books, 1964), p. 535.

37. In his discussion of the Renaissance, and particularly of Leonardo, Cassirer comments that "the focal point of intellectual life must lie in the place where the 'idea' is embodied, i.e., where the non-sensible form present in the mind of the artist breaks forth into the world of the visible and becomes realized in it" (*Individual and Cosmos,* p. 67). I take it that this is just what linear perspective permits, in Leonardo's case.

38. In contrast, Cassirer observes that "for Galileo, the homogeneity of the world follows from the necessary homogeneity of geometrical space" (*Individual and Cosmos,* p. 183).

39. Edgerton, *Renaissance Rediscovery,* p. 161.

40. "It is fair to say that without this conjunction of perspective and printing in the Renaissance, the whole subsequent development of modern science and technology would have been unthinkable" (Edgerton, *Renaissance Rediscovery,* p. 164).

41. Cassirer, *Individual and Cosmos,* p. 158.

42. For a discussion of the role of vision in the formulation of the contemporary understanding of nature, see Evernden, *The Natural Alien,* pp. 87 ff.

43. See Michael Baxandall, *Painting and Experience in Fifteenth*

Century Italy, 2d ed. (Oxford: Oxford University Press, 1988), p. 87.

44. On the emergence of the literal, see Owen Barfield's discussion in *Saving the Appearances* (New York: Harcourt Brace & World, 1965), esp. chap. 13.

Chapter 5: The Literal Landscape

1. See, e.g., E. A. Burtt's discussion in *The Metaphysical Foundations of Modern Science*, p. 79. On the contrast between the Italian preference for the ideal and the northern artists' attention to the particular, see Svetlana Alpers, *The Art of Describing* (Chicago: University of Chicago Press, 1983), p. 78:

> One could say that Italian art was based on an intentional turning away from individuality in the name of general human traits and general truths. In such an art *resemblance* to certain ideals of appearance or of action, and thus resemblance between things, was constitutive of truth. This not only helped give the art a certain ideal cast, it also encouraged the differentiation between kinds of works. Portraiture, since it must attend to individuals, was considered inferior to works that engaged higher, more general human truths. The Dutch trust to and privileging of portraiture, which is at the center of their entire pictorial tradition, is connected on the other hand to a desire to preserve the identity of each person and each thing in the world.

2. In the initial discussion of this question of apparently superfluous detail, Bryson is elaborating on themes put forth by semiologists following in the tradition of Ferdinand de Saussure, such as Roland Barthes, whose work has been referred to earlier. Bryson's own discussion goes well beyond that, and entails certain challenges to Barthes's work. In particular, he is critical of the assumption that the codes of meaning must inevitably arise solely from *within* a semiological system, when there may well be additional sources of interpretation beyond it. See *Vision and Painting*, particularly chap. 4.

3. Ibid., p. 60.

4. Hence, says Bryson, "in proportion as denotation recedes, the knowledge drawn upon by the viewer enters into ever closer contact with the contextuality where alone it manifests" (ibid., pp. 74–75).

5. Ibid., p. 89. See Bryson's discussion of "the disavowal of deictic reference" in chap. 5 (pp. 89 ff.), in which he speaks of the use of oil painting as an "erasive medium" to permit the obscuring of marks of authorship.

6. Svetlana Alpers suggests that "instead of doing a traditional portrait of himself at work, or including himself in the form of a portrait, Vermeer disappears into the very act of observing and painting" (*Art of Describing*, p. 106).

7. Bryson, *Vision and Painting*, p. 107.

8. Alpers, *Art of Describing,* p. 45.

9. Alpers says of Dutch painting that "the attitude is conditional on a double fragmentation: first, the viewer's eye is isolated from the rest of his body at the lens; second, what is seen is detached from the rest of the object and from the rest of the world" (ibid., p. 85).

10. Cassirer, *Individual and Cosmos,* p. 154.

11. See Alpers, *Art of Describing,* p. 78.

12. Cited by Alpers, ibid., p. 223. Incidentally, given his general opposition to the upsurge of empiricism, it is not surprising to find William Blake seconding this opinion of Dutch art. He begins his poem, "Florentine Ingratitude"—largely an attack on Sir Joshua Reynolds and all that he represented—thus:

Sir Joshua sent his own Portrait to
the birth Place of Michael Angelo,
And in the hand of the simpering fool
He put a dirty paper scroll,
And on the paper, to be polite,
Did "Sketches by Michael Angelo" write.
The Florentines said, "'Tis a Dutch English bore,
"Michael Angelo's Name writ on Rembrandt's door."
The Florentines call it an English Fetch,
For Michael Angelo did never sketch.
Every line of his has Meaning
And needs neither Suckling nor Weaning.
'Tis the trading English Venetian cant
To speak Michael Angelo & Act Rembrandt.

13. Bryson, *Vison and Painting,* p. 60.

14. Cited in Alpers, *Art of Describing,* p. 91.

15. Nor can he seem to decide, at times, whether reason (through mathematics) or experience should form the foundation of our knowledge of nature. See Cassirer, *Individual and Cosmos,* p. 154.

16. Alpers, *Art of Describing,* p. 48.

17. In Jung's understanding, this would be temperamentally impossible (assuming Leonardo's introvert status), for "a series of concrete representations conditioned by sensuous perceptions is not exactly what the abstract thinker would call thinking but at best only passive apperception" (*Psychological Types,* p. 303).

18. Alpers, *Art of Describing,* p. xxv.

19. Shapin and Schaffer, *Leviathan and the Air-Pump,* pp. 17–18.

20. Ibid., p. 23.

21. Ibid., p. 62.

22. Alfred North Whitehead, in his *Science and the Modern World* (New York: Free Press, 1967), suggested that even though empiricists "deserted the method of unrelieved rationalism," their approach nevertheless entails reason: "We must observe the immediate occasion, and *use reason* to elicit a general description of its nature. Induction

presupposes metaphysics. In other words, it rests upon an antecedent rationalism" (pp. 43, 44).

23. It may well be shadows of this debate that we see in the ongoing controversies in ecology between the "historical" and "ahistorical" approaches. See Sharon Kingsland, *Modelling Nature: Episodes in the History of Population Ecology* (Chicago: University of Chicago Press, 1985), or Robert McIntosh, *The Background of Ecology* (Cambridge: Cambridge University Press, 1985).

24. This is a somewhat misleading assertion in that it seems to suggest that a mode of investigation, commonly called empiricism, achieved dominance in the seventeenth century and has gone unchallenged to the present. This is surely not the case. However, in deliberating on whether to elaborate on later modifications, I decided that it would do little to advance the intentions of this essay and might indeed muddy the already turbid waters that we are probing. I think it is fair to say that our modern presuppositions still tend to be those of the empiricist, even though it may be, as Michel Foucault has argued, that there has been a significant shift in attention toward "the new empiricities." Essentially, Foucault suggests that from an attention to visible surfaces through which organisms can be ordered and classified, there is a shift in interest toward the functional, that is, toward the systemic operation of the organism in its invisible core. "When we consider the organ in relation to its function, we see, therefore, the emergence of 'resemblances' where there is no 'identical' element; a resemblance that is constituted by the transition of the function into evident invisibility." This allows us to speak of these invisible functionalities as if they were actual things, and "life" emerges as a topic: "Up to the end of the eighteenth century, in fact, life does not exist: only living beings" (*The Order of Things*, pp. 250, 264, 160).

If Foucault is correct, this is certainly a significant change. Indeed it may, as Charles Bergman has suggested in *Wild Echoes* (Anchorage: Alaska Northwest Books, 1990), even have implications for the treatment of endangered species, through the very definition of "animal" which comes to be articulated. However, in the short run at least, Foucault's refinement does not seem to add an essential dimension to the discussion of "Nature" that has been developed here (although it might be regarded as yet another "oscillation" between the empathetic and the abstract descriptions of nature). For our purposes, the changes of significance remain the disengagement of the human from the natural world, the maintenance of that separation through the ban on empathetic "projection," and the gradual elaboration of a domain called Nature which is reasonable at the core, and hence open to prediction and control. It is that fundamental dualism, along with the expectation of increasingly accurate representations of Nature and unlimited license to manipulate it, that facilitates the contemporary treatment of nonhuman others (and doubtless of other humans as well). Attention to the "new empiricities" may exacerbate that treatment but does not fundamentally alter it.

25. For an excellent treatment of the significance of Francis Bacon in the development of attitudes toward nature, see William Leiss's *The Domination of Nature* (Boston: Beacon Press, 1974), a book that seems to have been published "before its time" and consequently has not received the attention it deserves.

26. See Shapin and Schaffer's excellent account in *Leviathan and the Air-Pump.*

27. Owen Barfield has suggested that the camera obscura was a significant agent in the transformation of thought, that "revolution, formulated rather than initiated by Immanuel Kant, whereby the human mind more or less reversed its conception of its own relation to its environment. It is more than an emblem, because the camera obscura (considered as the original source of the whole camera sequence) was also *instrumental* in actually bringing about the change of which I have spoken." He suggests that the camera has become our "ruling metaphor" (*The Rediscovery of Meaning* [Middletown: Wesleyan University Press, 1977], p. 69).

28. See Alpers, *Art of Describing,* p. 36.

29. See Alpers's discussion of the new model of visual perception that accompanies this shift to empiricism (ibid., pp. 33 ff.).

30. William Ivins, Jr., *The Rationalization of Sight* (New York: Da Capo, 1973), p. 13.

31. Jonas, *Phenomenon of Life,* p. 152.

32. On this, see Rorty's discussion of Locke in *Philosophy and the Mirror of Nature,* pp. 140 ff.

33. Northrop Frye, *The Great Code* (Markham: Penguin Books, 1990), pp. 7–8.

34. Ibid, p. 13.

35. Of course, even for the scientists, there was a problem in attaining knowledge through vision, for there was always the possibility of error. In effect, they proposed a kind of *ideal* vision, an idea that "if we could only do this"—such as stand on the moon and look down—"then we would truly see that." Frye concludes that "the problem of illusion and reality . . . becomes a central one in third-phase language" (ibid., p. 14).

36. Michel Foucault suggests that from the seventeenth century, "similitude is no longer the form of knowledge but rather the occasion of error," and that "without imagination, there would be no resemblance between things" (*The Order of Things,* pp. 51, 69). Foucault probably does not mean by "imagination" what Blake did, but the change he speaks of—the abandonment of the search for underlying similarities in favor of the investigation of distinctions—is probably the event that provoked Blake's outrage.

37. Indeed, the Romantic movement and its descendants in many ways epitomize this rejection of the "mirror" or "camera" model of knowledge. See M. H. Abrams, *The Mirror and the Lamp* (Oxford: Oxford University Press, 1971), or his later book, *Natural Supernaturalism* (New York: Norton, 1973), for detailed discussion.

38. Jung, *Psychological Types,* p. 318.

39. Because he feels there are many varieties of rationalism, Jung prefers to shun the term in favor of "idealism" (ibid., pp. 307–8).

40. Ibid., p. 310.

41. Ibid., p. 311.

42. C. S. Lewis, *The Discarded Image,* p. 206.

43. This has been discussed particularly in relation to Francis Bacon's admonitions against attending to overarching similarities rather than to particulars. See, e.g., Foucault, *The Order of Things,* pp. 51–52:

> We already find a critique of resemblance in Bacon—an empirical critique that concerns, not the relations of order and equality between things, but the types of mind and the forms of illusion to which they might be subject. We are dealing with a doctrine of the *quid pro quo.* Bacon does not dissipate similitudes by means of evidence and its attendant rules. He shows them, shimmering before our eyes, vanishing as one draws near, then re-forming again a moment later, a little further off. They are *idols.* The *idols of the den* and the *idols of the theatre* make us believe that things resemble what we have learned and the theories we have formed for ourselves; the other idols make us believe that things are linked by resemblances between ourselves.

44. Frye, *The Great Code,* p. 13.

45. See Lewis, *Discarded Image,* p. 204.

46. Jonas, *Phenomenon of Life,* pp. 192–93.

47. Ibid., p. 190.

48. Ibid., p. 201.

49. Owen Barfield, *Poetic Diction,* 3d ed. (Middletown: Wesleyan University Press, 1973), p. 24.

50. Cassirer, *Individual and Cosmos,* p. 134.

51. Barfield, *Poetic Diction,* p. 32.

52. Parenthetically, it is this provision that E. F. Schumacher criticized in his discussion of "adequacy" on the part of observers in his *A Guide for the Perplexed* (New York: Harper Colophon, 1978).

53. Northrop Frye, *Fearful Symmetry: A Study of William Blake* (Princeton: Princeton University Press, 1969), pp. 21–22.

54. Frye, *The Great Code,* p. 21.

55. Van den Berg, *The Changing Nature of Man,* p. 207.

56. Ibid., p. 217.

57. Ibid., p. 230.

58. Again, it is just this that rebels like William Blake railed

against. Blake claimed that when he saw the sun, it was not a round disc, a "guinea sun," but rather "an Innumerable company of the Heavenly host crying, 'Holy, Holy, Holy is the Lord God Almighty.'" In *Fearful Symmetry*, p. 21, Northrop Frye asks:

> The Hallelujah-Chorus perception of the sun makes it a far more real sun than the guinea-sun, because more imagination has gone into perceiving it. Why, then, should intelligent men reject its reality? Because they hope that in the guinea-sun they will find their least common denominator and arrive at a common agreement which will point the way to a reality about the sun independent of their perception of it. The guinea-sun is a sensation assimilated to a general, impersonal, abstract idea. Blake can see it if he wants to, but when he sees the angels, he is not seeing more "in" the sun but more of it.

59. As van den Berg observes, "The factualization of our understanding—the impoverishment of things to a uniform substantiality—and the disposal of everything that is not identical with this substantiality into the 'inner self' are both parts of one occurrence. The inner self became necessary when contacts were devalued" (*Changing Nature*, p. 227).

Chapter 6: The Fragile Division

1. There is much that should be said about the Greek experience, but to do so would be to compound the complexity of our topic. It must suffice to remind the reader of the enormous significance of the Greek transformation from a "mythic" to a "theoretic" society, a transformation which is certainly connected with the discernment of "nature" and with the subsequent ability of their intellectual descendants, including ourselves, to treat nature as a particular kind of object or set of objects. Indeed it has been suggested that the very attention to objects or things only becomes possible at this juncture, when they can be construed as separate from their viewers and become open to our wonder and our speculation. Julian Marias suggests that "theorizing consciousness . . . sees *things* where previously it saw only powers. This constitutes the great discovery of *things*, a discovery so profound that today it is difficult for us to realize that it actually was a discovery or to imagine that it could have happened any other way" (*History of Philosophy* [New York: Dover, 1967], p. 4). Perhaps the most profound discussion of the ascendancy of interest in "beings" rather than in "Being" can be found in the work of Martin Heidegger, particularly his early writing, *Being and Time* (New York: Harper & Row, 1962).

2. This example derives from Jakob von Uexkull's explication of the subjective worlds or "umwelts" of animals. His work was unique in its day, and still commands considerable interest. See, e.g., Thomas

A. Sebeok's discussion in *The Sign and Its Masters* (Lanham, Md.: University Press of America, 1989), pp. 187–207.

3. Jonas, *Phenomenon of Life*, p. 12.

4. Ibid., pp. 13–14.

5. The mind-body problem is, of course, discussed in great detail in most philosophy texts. Being a poet, however, Dennis Lee provides a far more graphic image of it than do most academics. His essay on cosmology, from which this example is drawn, rewards careful reading and contains a highly original vision of the relationship of World and Earth. See Dennis Lee, *Savage Fields: An Essay in Literature and Cosmology* (Toronto: Anansi, 1977).

6. Ibid., pp. 52–53.

7. Ibid., p. 53.

8. Lewis, *Discarded Image*, pp. 214–15.

9. In *Phenomenon of Life*, p. 18, Jonas writes:

> Dualism, when its work was done, had left behind the "extended" as the lifeless and unfeeling, and the body undeniably is a part of this extended: either, then, it is essentially the same as the extended in general—then its being alive is not understood; or it is *sui generis*—then the exception claimed for it is not to be understood and calls into question the whole rule.

10. The question of our means of knowing, along with the consequences of our overriding concern with dualism and epistemology, is dealt with by many philosophers, but perhaps by none more successfully than Richard Rorty in *Philosophy and the Mirror of Nature*.

11. On this, see Christopher Lasch, *The True and Only Heaven: Progress and its Critics* (New York: Norton, 1991).

12. Raymond S. Stites, M. Elizabeth Stites, and Pierina Castiglione, *The Sublimations of Leonardo da Vinci* (Washington, D.C.: Smithsonian Institution Press, 1970), p. 367.

13. Jonas, *Phenomenon of Life*, p. 16. He continues (p. 17):

> The problem is still the same: the existence of feeling life in an unfeeling world of matter which in death triumphs over it. If its dualistic solution is theoretically unsatisfactory, the two partial monisms—materialism and idealism—at bottom evade it, each in its own manner of one-sidedness. Their means of unification, i.e., of reduction to the chosen denominator, is the distinction of primary and secondary reality: of substance and function (or "epiphenomenalism") in the case of materialism, of consciousness and appearance in the case of idealism.

14. Ibid., p. 20.

15. Joseph Campbell, *Myths to Live By* (New York: Bantam, 1973), pp. 1–2.

16. C. S. Lewis, *The Abolition of Man* (New York: Macmillan, 1955), p. 83.

17. On the question of the "natural," see Loren Eiseley, *The Firmament of Time* (New York: Atheneum, 1972).

18. For a discussion of the relation of scientific agriculture to ecology, see Donald Worster, *Nature's Economy* (Garden City: Doubleday Anchor, 1979).

19. See John Livingston, "Rightness or Rights?" *Osgoode Hall Law Journal* 22, 2 (Summer 1984): 309–21.

20. See William Barrett, *Irrational Man: A Study in Existential Philosophy* (Garden City: Doubleday, 1962), pp. 217 ff. See also my discussion of "fields of self" in *The Natural Alien.*

21. This position is elaborated in much of the "deep ecology" literature. See, e.g., Bill Devall and George Sessions, *Deep Ecology: Living as if Nature Mattered* (Salt Lake City: Peregrine Smith Books, 1985), or Warwick Fox, *Toward a Transpersonal Ecology: Developing New Foundations for Environmentalism* (Boston: Shambhala Publications, 1990).

22. See my "Nature in Industrial Society," in Ian Angus and Sut Jhally, eds., *Cultural Politics in Contemporary America* (New York: Routledge, 1989), pp. 151–66. A somewhat similar contrasting of two "natures" can be found in Peter Reed's "Man Apart: An Alternative to the Self-Realization Approach," *Environmental Ethics* 11, 1 (Spring 1989): 53–69.

23. Alain Robbe-Grillet, "Dehumanizing Nature," in Richard Ellmann and Charles Feidelson, eds., *The Modern Tradition* (New York: Oxford University Press, 1965), pp. 361–78.

24. Michael Oakeshott's "The Voice of Poetry in the Conversation of Mankind," in *Rationalism in Politics* (New York: Basic Books, 1962), pp. 197–247, contains a discussion that is especially relevant to this point.

Chapter 7: Nature and the Ultrahuman

1. Keith Thomas, *Man and the Natural World: Changing Attitudes in England, 1500–1800* (Harmondsworth: Penguin Books, 1984), p. 89.

2. Marjorie Grene, *Approaches to a Philosophical Biology* (New York: Basic Books, 1968), pp. 12–13. Grene's book, and particularly her initial essay on Adolph Portmann, is especially relevant to the question of what we regard as "nature" and to the general discussion of this chapter. On the question of the constitution of our "lifeworld," see Evernden, *The Natural Alien.*

3. John Berger, *About Looking* (New York: Pantheon, 1980), p. 14. See also Charles Siebert, "Where Have All the Animals Gone?" *Harper's* 282 (May 1991): 49–58.

4. Ibid., pp. 13 ff.

5. Barfield, *Poetic Diction,* p. 32.

6. Erazim Kohak, *Idea and Experience* (Chicago: University of Chicago Press, 1978), p. 11.

7. Ibid., p. 33.

8. Maurice Merleau-Ponty, *Phenomenology of Perception* (London: Routledge & Kegan Paul, 1962), p. ix.

9. In *Phenomenology of Perception,* p. 24, Merleau-Ponty commented:

> The nature about which empiricism talks is a collection of stimuli and qualities, and it is ridiculous to pretend that nature thus conceived is, even in intention merely, the primary object of our perception: it does in fact follow the experience of cultural objects, or rather is one of them. We shall, therefore, have to rediscover the natural world too, and its mode of existence, which is not to be confused with that of the scientific object.

10. Marshall Sahlins has admirably discussed the social utility of nature. He notes, for example, that not all cultures make the same color distinctions, or attribute identical meanings to colors. But all have and use these distinctions in some way—it is a natural feature that facilitates human signification. In *The Use and Abuse of Biology* (London: Tavistock, 1977), p. 66, Sahlins writes:

> It is exactly because colors subserve this cultural significance that only certain precepts biologically available to human beings become "basic," namely those that by their distinctive contrasts and perceptual relations, such as uniqueness or complementarity of hue, can function as signifiers in meaningful systems.

Colors become not simply perceptual givens, but signifiers of meaning: they become *ours,* just as the grand artifact, Nature, has also become ours. Again, nature is a product of culture. In *Culture and Practical Reason* (Chicago: University of Chicago Press, 1976), p. 209, Sahlins observes:

> For all their facticity and objectivity, the laws of nature stand to the order of culture as the abstract to the concrete: as the realm of possibility stands to the realm of necessity, as the given potentialities to the one realization, as survival is to actual being. That is because nature is to culture as the constituted is to the constituting. Culture is not merely nature expressed in another form. Rather the reverse: the action of nature unfolds in the terms of culture; that is, in a form no longer its own but embodied as meaning.

11. In his *Phenomenology of Perception,* p. 57, Merleau-Ponty comments:

> The first philosophical act would appear to be to return to the world of actual experience which is prior to the objective world, since it is in it that we shall be able to grasp the theoretical basis no less than the limits of the objective world, restore to things their concrete physiognomy, to organisms their individual ways of

dealing with the world, and to subjectivity its inherence in history. Our task will be, moreover, to rediscover phenomena, the layer of living experience through which other people and things are first given to us, the system "Self-others-things" as it comes into being; to reawaken perception and foil its trick of allowing us to forget it as a fact and as perception in the interest of the object which it presents to us and the rational tradition to which it gives rise.

Erazim Kohak's *The Embers and the Stars* (Chicago: University of Chicago Press, 1984) is a poignant illustration of this endeavor.

12. I do not mean to suggest that the child is a "blank slate," or that a completely naive encounter with the world is necessarily involved. However, the child does at least encounter the world with a minimum of societal censorship, and probably comes as close as a person can to direct and uncontrolled encounter with otherness. See Richard N. Coe, *When the Grass Was Taller: Autobiography and the Experience of Childhood* (New Haven: Yale University Press, 1984), particularly chap. 3.

13. This is certainly not an uncommon endeavor. It is well illustrated in Thomas Carlyle's *Heroes, Hero Worship, and the Heroic in History* (New York: A. L. Burt, 1892), pp. 8–10. He builds his image on Plato's story of the cave, and suggests that to such a person "simple, open as a child," nature would have "as yet no name" and would be "*preter*natural":

> This green flowery rock-built earth, the trees, the mountains, rivers, many-sounding seas; that great deep sea of azure that swims overhead; the winds sweeping through it; the black cloud fashioning itself together, now pouring out fire, now hail and rain; what *is* it? Ay, what? At bottom, we do not yet know; we can never know at all. It is not by our superior insight that we escape the difficulty; it is by our superior levity, our inattention, our *want* of insight. It is by *not* thinking that we cease to wonder at it. Hardened round us, encasing wholly every notion we form, is a wrappage of traditions, hearsays, mere *words*. . . . This world, after all our science and sciences, is still a miracle; wonderful, inscrutable, *magical* and more, to whosoever will *think* of it.

14. See William Barrett, *Irrational Man: A Study in Existential Philosophy* (Garden City: Doubleday, 1962), esp. pp. 218 ff.

15. For the sake of simplicity, I am speaking as if it is only the infant we are concerned with. Possibly—in fact probably—this discovery of Otherness occurs more than once and in different ways. There may well be an equally important encounter in the middle ages of childhood, and again in adolescence. This possibility has been eloquently described by Paul Shepard, most notably in his *Nature and Madness* (San Francisco: Sierra Club Books, 1982) and *Thinking Animals* (New York: Viking, 1978).

16. Gaston Bachelard, *The Poetics of Space* (Boston: Beacon Press, 1969), p. 156.

17. Kathleen Raine, *Farewell Happy Fields* (New York: Braziller, 1977), p. 12.

18. In dealing with the complex works of Martin Heidegger, George Steiner (*Heidegger* [Glasgow: Fontana/Collins, 1978], pp. 68–69) offered a description which may help illuminate the present question of encountering "radical otherness." He said:

> Being is not an entity in itself . . . which generates the powers of manifestation in beings. It is hidden Being that gives the rock its dense "thereness," that makes the heart pause when a kingfisher alights, that makes our own existence inseparable from that of all others. In each case, wonder and reflection tell us of an intensity of presentness, of an integral unfolding of self-statement, *clearly in excess of sensory data and neutral registration*.

I do not wish to make too much of this comparison, or to suggest a deliberate Heideggerian undercurrent in our discussion. I raise this comparison only to underline the point that the encounter with radical otherness always implies something *beyond* the ordinary, beyond that which can be readily assigned to a conceptual category or distinguished as simply one thing among others. There is always a sense of significance that transcends ready comprehension, and of a revelation of something for which there is no adequate language.

Richard Coe has also suggested that this kind of realization functions in the childhood experience of nature. In his survey of autobiographies, Coe finds that despite a gradual diminishment of attachment to formal religions, the sense of mystery does not appear to diminish. In fact, this change has led "to a world more obsessively mysterious than ever, since the abandoning of overfacile explanations has left man face to face with the ultimately inexplicable: the 'facticity' of the world, the superfluousness of pure existence" (*When the Grass Was Taller*, p. 116).

19. Edith Cobb, *The Ecology of Imagination in Childhood* (New York: Columbia University Press, 1977), p. 30.

20. Ibid., p. 32.

21. See Anthony Storr, *Solitude: A Return to the Self* (New York: Free Press, 1988).

22. Not unlike, perhaps, William Blake's contention in *Jerusalem* that "'I must Create a System or be enslav'd by another Man's / I will not Reason & Compare: my business is to Create' " (Geoffrey Keynes, ed., *Poetry and Prose of William Blake* [London: Nonesuch, 1967], p. 442). For further discussion of the Romantic view that the artist must create a world, see M. H. Abrams, *Natural Supernaturalism* (New York: Norton, 1971), p. 256.

23. Cobb felt that the age at which this was most important was

between six and twelve years, but Paul Shepard, in his *Nature and Madness,* places the important stage much earlier.

24. Cobb, *Ecology of Imagination,* pp. 28–29.

25. Coe, *When the Grass Was Taller,* p. 120.

26. Gaston Bachelard, *Water and Dreams: An Essay on the Imagination of Water* (Dallas: Pegasus Foundation, 1983), p. 20.

27. Jung, *Psychological Types,* p. 312.

28. Merleau-Ponty, *Phenomenology of Perception,* p. xiv.

29. Ibid., p. xvi.

30. Bachelard, *Water and Dreams,* p. 1.

31. Ivan Illich, *H_2O and the Waters of Forgetfulness* (Berkeley: Heyday Books, 1985), p. 76. There is also an interesting use of the H_2O/water contrast in Michael Oakeshott's "The Voice of Poetry in the Conversation of Mankind" (*Rationalism in Politics and Other Essays* [New York: Basic Books, 1962], p. 222):

> Every mode of imagining is activity in partnership with images of a specific character which cannot appear in any other universe of discourse. . . . For example, the word "water" stands for a practical image; but a scientist does not first perceive "water" and then resolve it into H_2O: *scientia* begins only when "water" has been left behind. To speak of H_2O as "the chemical formula for water" is to speak in a confused manner: H_2O is a symbol the rules of whose behavior are wholly different from those which govern the symbol "water." And similarly, contemplative activity is never the "conversion" of a practical or a scientific image into a contemplated image; its appearance is possible only when practical and scientific imagining have lost their authority.

32. The idea of "laws" of nature is curiously anthropomorphic, conjuring images of natural objects fearing pursuit by some cosmic police force should they breach the legislated code of behavior. This is pointed out by C. S. Lewis, *The Discarded Image,* p. 93, as well as by Northrop Frye, *The Great Code,* p. 16. J. H. van den Berg entertains a wider consideration of these laws and of the conditions for their applicability in *The Changing Nature of Man,* pp. 121–25.

33. Rudolf Otto, *The Idea of the Holy* (London: Oxford University Press, 1950), p. 26. For a different application of Otto's ideas to the question of human/nature interaction, see Peter Reed, "Man Apart," pp. 53–69. Reed argues that, in contrast to the assumptions of the "deep ecology" movement, the realization of radical dissimilarity between humans and nature might have positive consequences. For a rejoinder to this paper, see Arne Naess, "Man Apart and Deep Ecology: A Reply to Reed," *Environmental Ethics* 12, 2 (Summer 1990): 185–92.

34. Otto, *Idea of the Holy,* p. 79.

35. Coe, *When the Grass Was Taller,* p. 135. The fact that it is

inexplicable is, strangely to us, of enormous value. The inexplicable is in short supply today. Perhaps it is an impossibility in our understanding of Nature. But it may be that our craving is nonetheless apparent, bearing in mind that "'the monstrous' is just the 'mysterious' in a gross form" (Otto, *Idea of the Holy,* p. 80). The fascination which the young display for "monster movies" and the like are understandable in light of the fact that it is their sole source of the inexplicable and the uncanny. Their education is devoted to rational explanation, and their religion, for those that have any, tends to bend over backwards to demonstrate how reasonable and literal-minded it can be. It may be that the only source of the wondrous and the strange, now extinguished in Nature, is Hollywood: a poor substitute. In *When the Grass Was Taller,* p. 116, Richard Coe further notes that we apparently have a better chance of

> penetrating the mystery of God than of grasping the unfathomable reality of a pebble on the seashore, for God, it may be assumed, is at least Mind, whereas the pebble is not-Mind. To apprehend the reality of that which is not-Mind is the severest challenge which the human intellect can encounter.

36. Howard L. Parsons, "A Philosophy of Wonder," *Philosophy and Phenomenological Research* 30 (1969–70): 84–101.

37. R. W. Hepburn, *"Wonder" and Other Essays* (Edinburgh: Edinburgh University Press, 1984), p. 134.

38. Ibid., p. 144.

39. John Fowles, "Seeing Nature Whole," *Harper's* 259 (November 1979): 66.

40. Interestingly, another poet who was obsessed with the encounter with Otherness coined a similar term to facilitate his discussion. Robinson Jeffers speaks of the "inhuman," and while many would regard this as further evidence of his allegedly misanthropic character, William Everson presents a rather more positive interpretation of Jeffers and his work in *The Excesses of God: Robinson Jeffers as a Religious Figure* (Stanford: Stanford University Press, 1988). He also interprets Jeffers's work in the light of Rudolph Otto's concept of the Holy.

41. Richard Jefferies, *Landscape with Figures: An Anthology of Richard Jefferies's Prose* (Harmondsworth: Penguin Books, 1983), p. 288.

42. Richard Jefferies, *The Story of My Heart* (London: Duckworth, 1923), p. 48.

43. Jefferies, *Landscape with Figures,* p. 244.

44. The question of "direct encounter" is a difficult one, and is approached in a variety of ways. For example, Whitehead says of Wordsworth that "it is the brooding presence of the hills which haunts him. His main theme is nature *in solido,* that is to say, he dwells on that mysterious presence of surrounding things, which im-

poses itself on any separate element that we set up as an individual for its own sake" (*Science in the Modern World*, p. 83). See also Owen Barfield on "concrete thinking": "The time-honoured 'subjective-objective' dichotomy vanishes in the light of concrete thinking; and the word *concrete* can perhaps best be defined as 'that which is neither objective nor subjective' " (*Poetic Diction*, p. 210).

45. In his essay "Dehumanizing Nature" (in Richard Ellman and Charles Feidelson, eds., *The Modern Tradition* [New York: Oxford University Press, 1965], p. 367), Alain Robbe-Grillet addresses this tendency quite bluntly:

> If I say, "The world is mankind," I shall always obtain absolution; but if I say, "Things are things, and man is only man," I shall be immediately judged guilty of a crime against humanity.
>
> The crime is to state that something exists in the world which is not mankind, which makes no signs to man, which has nothing in common with him. From their viewpoint, the crime lies especially in recognizing this separation, this distance, without making any effort to transcend or to sublimate it.

46. It is perhaps not surprising that Peter Reed concluded that "the danger is not only in physical interferences with nature. Nature's wonder also recoils under the onslaught of our mental models and our sciences. The awe of the Numen cannot coexist with the drabness of the known" ("Man Apart," p. 64).

47. Fowles, "Seeing Nature Whole," p. 67.

48. Henry David Thoreau, *The Writings of Henry David Thoreau*, vol. 5, *Excursions and Poems* (New York: AMS Press 1968), p. 224.

49. See Thomas H. Birch, "The Incarceration of Wildness: Wilderness Areas as Prisons," *Environmental Ethics* 12, 1 (Spring 1990): 3–26. He observes (p. 22) that "managing wildness is contradictory" for "wildness is logically intractable to systemization. There can be no natural laws of wildness."

50. Annie Dillard, *Teaching a Stone to Talk: Expeditions and Encounters* (New York: Harper Colophon, 1983), pp. 14–15.

51. Barfield, *Poetic Diction*, p. 177.

52. Robbe-Grillet, "Dehumanizing Nature," p. 363.

53. Bachelard, *Water Dreams*, p. 16.

54. Richard Jefferies, *Landscape with Figures*, p. 238.

Epilogue

1. Jefferies, *The Story of My Heart*, pp. 98–99. As noted earlier, the idea of "chaos" as an appropriate descriptor of nature has achieved some prominence in ecological circles, although it is by no means clear at this stage what the implications of this may be. See Donald Worster, "The Ecology of Order and Chaos," *Environmental History Review* 14 (1990): 1–18.

2. C. S. Lewis makes the interesting observation that *genius* is the standard Latin translation of *daemon,* and that the individual *daemon* was thought in the Middle Ages to be allotted to each person as a "witness or guardian." In *The Discarded Image,* p. 42, Lewis notes that through a tortuous transition,

> a man's *genius,* from being an invisible, personal, and external attendant, became his true self, and then his cast of mind, and finally (among the Romantics) his literary or artistic gifts. To understand this process fully would be to grasp that grand movement of internalisation, and that consequent aggrandisement of man and desiccation of the outer universe, in which the psychological history of the West has so largely consisted.

It would appear, then, that "demons" were expunged in more than one sense by Leonardo and his successors.

3. Charles Bergman, *Wild Echoes: Encounters with the Most Endangered Animals in North America* (Anchorage: Alaska Northwest Books, 1990). In considering knowledge and power, he is specifically reacting to arguments put forward by Michel Foucault (see, e.g., pp. 80–81).

4. Michael Oakeshott has suggested that the voices which dominate our conversation today are those of science and practical activity. In themselves, these are useful idioms; as holders of the conversational monopoly, they pose a danger, for "an established monopoly will not only make it difficult for another voice to be heard, but it will also make it seem proper that it should not be heard: it is convicted in advance of irrelevance" ("The Voice of Poetry in the Conversation of Mankind," in *Rationalism in Politics and Other Essays* [New York: Basic Books, 1962], p. 202).

5. Bergman, *Wild Echoes,* p. 79.

6. Gifford Pinchot, *The Fight for Conservation* (Seattle: University of Washington Press, 1967), p. 45.

7. Fro Harlem Brundtland, *Our Common Future/World Commission on Environment and Development* (Oxford: Oxford University Press, 1987).

8. Charles Bergman puts this point more strongly: "Unless we can see them, unless we can count them, unless we can put numbers to their population and grant them this empirical existence, they are somehow not real to us" (*Wild Echoes,* p. 211). This requirement that the beast be firmly located in physical space has a corollary: that it not exist *in us.* In consigning endangered species to strong forms of control and captivity, we have found a way "to confirm our alienation from the beast—quite literally we ensure that it is no longer *inside* us but living in a separate designated space, a beast we have isolated" (p. 82).

To control animals by designating the way in which they may be known and spoken of is a form of self-domestication, a banishing of one's own unruly wildness and bestiality. It also denies the animal the right to dwell in us, so to speak, and in so doing undermines the kinship of the imagination which is prerequisite to sympathy and solidarity. Bergman speaks of Keats's notion of "negative capability," that is, "the power of the imagination to enter into another being, to participate in the immediate experience of nature," and suggests that "the imagination offers another ark, where animals can be reborn inside of us" (pp. 53, 114). It is perhaps not too great a stretch to suggest that the "imaginative identification" which Rorty suggests is necessary for the realization of human solidarity, the sense that we all share the capacity to suffer, is the very capacity that Bergman is alluding to, and which is banished from the vocabulary of conservation.

Indeed, it may be that the imposition of such a vocabulary is the only way in which we could *prevent* that "imaginative identification" which permits, even requires, that we "feel" the suffering of others. That is, if our means of speaking of nonhumans is such that identification is difficult or impossible—if we are only able to speak of them as objects or artifacts because the applicable vocabulary prohibits their subjectivity—then the possibility of our "noticing" their suffering is greatly diminished. (On imaginative identification and human solidarity, see Rorty, *Contingency, Irony, and Solidarity,* pp. 93–94.) Our mode of knowing and speaking extracts the otherness within and consigns it to the stockyard of the empiricist, thereby freeing us from the obligations to avoid inflicting pain and suffering on those with whom we share that capacity. It also, therefore, excludes the concept of limits from our conversation.

9. Thoreau, "Walking," in *The Writings of Henry David Thoreau,* 5:234.

10. Lewis, *The Abolition of Man,* pp. 82–83. In *Wild Echoes,* p. 7, Charles Bergman has made a similar observation in the contemporary context:

> Built into the question "Why save endangered species?" is all the arrogance of centuries of Western domination over nature. The question presupposes that we are the lords of creation, that it is our right and our duty to oversee nature. It also exposes our limited view of nature, even in our concern for it: Animals are something for us to control.

11. Walker Percy, *Lost in the Cosmos: The Last Self-Help Book* (New York: Washington Square Press, 1983), p. 105.

12. Rorty, *Philosophy and the Mirror of Nature,* p. 352.

13. The stress on the unfamiliar runs throughout Rorty's later discussions concerning metaphor, and although he is not explicitly concerned with the nonhuman world, he does seem to diminish boundaries and to employ references to that world in elaborating the significance of metaphor (particularly as it is discussed in the work of Donald Davidson). For example, he claims that "we can say that we

come to understand metaphors in the same way that we come to understand anomalous natural phenomena. . . . We interpret metaphors in the same sense in which we interpret such anomalies—by casting around for possible revisions in our theories which may help to handle the surprises." It is these surprises, these "catastrophes," which *reveal* the need for a vocabulary that can acknowledge them, and in which a different world can be spoken.

Later, in pointing out that Davidson attends to the shock-value of metaphor rather than to any alleged meaning, Rorty notes that metaphors are like surprising natural phenomena in that while they

> do not (literally) *tell* us anything . . . they do make us notice things and start looking around for analogies and similarities. They do not have cognitive content, but they are responsible for a lot of cognitions. For if they had not turned up we should not have been moved to formulate and deploy certain sentences which do have such content. As with platypuses, so with metaphors.

(*Objectivity, Relativism, and Truth* [Cambridge: Cambridge University Press, 1991], p. 167).

I suspect that Rorty's work has much to offer in the elaboration of "nature" and in the means to transformation that is sought by those who would, in any sense, seek to defend Earth. His potential significance in this domain has not gone unnoticed. Max Oelschlaeger, in his benchmark study of *The Idea of Wilderness,* notes that Rorty's idea of the role of "edifying philosophers"—to protest against attempts to thwart important conversation by entrenching a "privileged set of descriptions" (including what a bird is, presumably)—is very much like the one played by important figures in environmental preservation, such as John Muir (p. 174). Further, Oelschlaeger suggests that even though Rorty largely avoids discussion of nonhumans, he "opens the door to recognition of non-human others through his discussion of cruelty and pain" (p. 451). I presume he is referring to Rorty's tendency, in *Contingency, Irony, and Solidarity,* to occasionally follow statements such as this: "The idea that we all have an overriding obligation to diminish cruelty, to make human beings equal in respect to their liability to suffering, seems to take for granted that there is something within human beings which deserves respect and protection quite independently of the language they speak. It suggests that a nonlinguistic ability, the ability to feel pain, is what is important, and that differences in vocabulary are much less important" (p. 88), with provocative embellishments such as this: "Pain is nonlinguistic: It is what we human beings have that ties us to the nonlanguage-using beasts" (p. 94). As Oelschlaeger notes, there seems little reason why his idea of "solidarity" based on such nonlinguistic features as susceptibility to suffering could not be extended to nonhumans as well (*Idea of Wilderness,* p. 319). Rorty seems unwilling to make that extension, however, and instead falls back on the common humanist defense that human uniqueness authorizes the establishment of arbitrary barriers to exclude other beings from consideration: humans must of course experience "a special kind of pain" because they are capable of being humiliated through the willful destruction of their "structures of language and belief" (*Contingency, Irony, and Solidarity,* p. 177). Interestingly, these comments arise in the course of Rorty's discussion of George Orwell, making it all but impossible to resist observing that while all animals are equal in suffering pain, some are still, apparently, more equal than others. For a thorough consideration of animal pain, see Bernard E. Rollin, *The Unheeded Cry: Animal Consciousness, Animal Pain, and Science* (Oxford: Oxford University Press, 1990).

14. The term is John Dewey's, cited in Rorty, *Philosophy and the Mirror of Nature,* p. 379. For a pertinent and very original discussion of the significance of "missing metaphors" to the accommodation of people's experience of nonhuman others, see Annabelle Sabloff, "Re-Ordering the Natural World: Human-Animal Relations in a Canadian Urban Setting," Master's thesis, York University, 1991.

Bibliographical Essay

Raymond Williams once commented that although he had written about such ideas as culture, society, individual, class, art, and tragedy, "difficult as all those ideas are, the idea of nature makes them seem comparatively simple" (see "Ideas of Nature," in Jonathan Benthall, *Ecology: The Shaping Enquiry* [London: Longman, 1972], p. 146). Inevitably, any treatment of this topic is bound to be inadequate, and the reader can only correct the deficiencies of this one by consulting others which complement it. To assist in that process, I have selected a core of readings that seem to me to provide a fairly congenial entry into the larger literature. To give this collection some semblance of order, I have divided it according to the three sections of the book, the first group dealing principally with the question of the variability of the concept of nature, the second with the historic development from the medieval period to the present, and the third with the search for wild "otherness."

The Ambiguity of Nature

First, there is a set of works that deal in a very broad way with the question of human attitudes toward the natural world. Although I have placed them in this section, they really speak to all aspects of our topic. One of the best known is Clarence Glacken's *Traces on the Rhodian Shore* (Berkeley: University of California Press, 1967), which is subtitled "nature and culture in Western thought from ancient times to the end of the eighteenth century." It is a monumental undertaking, almost encyclopedic in scope. However, for that very reason, many readers may find that they can best use it by referring to specific topics (all well indicated in the table of contents and the index) rather than attempting to read it straight through from cover to cover.

Glacken's work provided some of the inspiration for Max Oelschlaeger's *The Idea of Wilderness: From Prehistory to the Age of Ecology* (New Haven: Yale University Press, 1991), which, although it very nearly matches Glacken's in scope, is a much more accessible work. Indeed, if I were forced to select a single source to recommend to someone seeking familiarity with "environmental thought," this would be the one. Although it covers a very wide time period, it manages to do so in surprising depth, and to treat both historic background and contemporary debate. It is a well-crafted book by a highly competent scholar.

On the issue of "environmentalism" and the debates surrounding it, the best discussion currently available is Warwick Fox's *Toward a Transpersonal Ecology* (Boston: Shambhala, 1990). While much of this may not relate to the question of "nature" explicitly, it does so indirectly because the environmental debate which it describes is, in large measure, a debate about how nature should be understood and what our relationship to it should be. Fox also provides a fine introduction to the eclectic literature of "environmental thought," which will certainly save anyone a lengthy and frustrating search for relevant readings.

Should the reader wish to start with shorter works of slightly narrower scope, R. G. Collingwood's *The Idea of Nature* (London: Oxford University Press, 1960) might suit. Collingwood's is a concise, philosophic treatment of the idea of nature from the Greeks, through the Renaissance, to the modern period. It provides quite explicit summaries, but perhaps not as much of the social context as one might wish. For that, perhaps only a work in the sociology of knowledge would suffice, and Peter Berger and Thomas Luckmann's *The Social Construction of Reality* (Garden City, N.Y.: Anchor Books, 1967) seems the obvious choice. The title really says it all, but the authors detail the process of construction with considerable skill, and in the process provide access to another range of potentially useful readings. Richard Rorty's *Philosophy and the Mirror of Nature* (Princeton: Princeton University Press, 1979) might be said to span both philosophy and the sociology of knowledge to some degree, but it is primarily a work of philosophy (particularly epistemology and hermeneutics) and will appeal mainly to readers of that persuasion. The insight he provides into our assumptions about "knowing nature" are extremely valuable, and Rorty's fluency makes him one of the more enjoyable philosophic writers.

Although very different in background and approach, Owen Barfield shares many of Rorty's interests and is probably more approachable for most readers. Barfield, a barrister by training and a companion of C. S. Lewis and J. R. R. Tolkien, has produced a series of fascinating and readable books over his long lifetime, most concerned with what he would call the "evolution of consciousness." The one that deals most specifically with questions of "nature" is *Saving the Appearances* (New York: Harcourt Brace & World, 1965). However, readers might also be interested in *Worlds Apart* (Middletown: Wesleyan University Press, 1971), a fictional conversation between people who represent different traditions of thought, or *The Rediscovery of Meaning and Other Essays* (Middletown: Wesleyan University Press, 1977), which contains a number of essays that elaborate on points raised in the current work (such as the question of the camera as the model for human perception, discussed in "The Harp and the Camera," pp. 65–78).

Not surprisingly, C. S. Lewis's interests overlap with Barfield's a good deal, but as a medieval scholar, Lewis naturally approaches them differently. I have relied on Lewis's discussion of the development of the concept of nature, partly because he has fleshed it out more fully than have most writers.

In almost all his work, there is some reference to the changing ideas of nature, and because of his familiarity with medieval thought, he is far better equipped than most to give us some glimpse of what it must have been like to see the world in that way. His *The Discarded Image* (Cambridge: Cambridge University Press, 1964) is one of the most accessible treatments of medieval thought, but he is also careful to relate his discussions to current ideas. *The Abolition of Man* (New York: Macmillan, 1955) is even more approachable and equally relevant—a point drawn to my attention some years ago by William Leiss, whose *The Domination of Nature* (Boston: Beacon Press, 1974) takes the story considerably further and does a marvelous job of describing the development of current attitudes toward nature.

Finally, I should mention two anthropologists, Mary Douglas and Marshall Sahlins. Since both are referred to in the text, it may seem unnecessary to add them to the current list, in which they seem somewhat out of place. I do so, however, for two reasons: first, because both are excellent and stimulating writers who would reward any reader's attention, and who most definitely do have something important to tell us about the idea of nature in our own society and beyond. Second, I include them here to acknowledge the very limited use I have made of the field of anthropology. I felt that to delve into the anthropological literature would take the book in quite a different direction than I intended, and would thus compromise the limited theme I was trying to develop. However, I would certainly encourage any reader interested in cross-cultural comparisons of the concept of nature to explore the anthropological literature, beginning with the two cited, or perhaps even with Paul Shepard, who has incorporated many anthropological insights into his important and original books (see, for example, *The Tender Carnivore and the Sacred Game* [New York: Scribner's, 1973]; *Thinking Animals* [New York: Viking, 1978]; *Nature and Madness* [San Francisco: Sierra Club Books, 1982]).

The Creation of Nature

Many of the sources cited above are, of course, also relevant to the question of the creation of nature, particularly C. S. Lewis, R. G. Collingwood, Owen Barfield, Max Oelschlaeger, and Clarence Glacken. To this list, however, we can add a number of others who, while not primarily occupied with the problem of nature, are extremely helpful in helping us understand the social context of the transformations that occurred.

First, on the understanding of the medieval outlook, there are several well-known works, including those of C. S. Lewis (cited above), J. Huizinga, *The Waning of the Middle Ages* (New York: Doubleday Anchor, 1954), David Knowles, *The Evolution of Medieval Thought* (New York: Vintage Books, 1962), and M.-D. Chenu, *Nature, Man, and Society in the Twelfth Century* (Chicago: University of Chicago Press, 1968). Also, works that cover a wider time period often have very good sections on both medieval and Renaissance thought: John Herman Randall's classic, *The Making of the Modern Mind* (Boston: Houghton Mifflin, 1940), remains one of the more approachable, and Arthur Lovejoy's *The Great Chain of Being* (New York: Harper & Row, 1960) is regarded by many as a benchmark in the history of ideas. While any of these can provide a good sense of the medieval period, it is a rare author who can give us a glimpse of what it might be like to actually see the world through medieval eyes. Owen Barfield does so on occasion (see *Saving the Appearances*), but one of the more successful attempts is Charles M. Radding's *A World Made by Men: Cognition and Society, 400-1200* (Chapel Hill: University of North Carolina Press, 1985). J. H. van den Berg also has some success in his *The Changing Nature of Man* (New York: Delta, 1975).

As will have become apparent, I find Ernst Cassirer one of the most interesting writers on the Renaissance, although many might consider him slightly passé today. Still, his stature is considerable, and his prose is clear. Much the same could be said of J. H. Randall, who is cited above, or of Hiram Haydn's *The Counter-Renaissance* (New York: Harcourt, Brace & World, 1950). There is one collection of essays, however, which has a somewhat special place in the literature of environmental affairs. I am speaking of Lynn White, Jr.'s *Dynamo and Virgin Reconsidered* (Cambridge: MIT Press, 1968), originally published as *Machina ex Deo: Essays in the Dynamism of Western Culture.* White's essays range widely, but the one that holds a special place in environmental thought is "The Historic Roots of our Ecologic Crisis," which caused considerable debate after its publication in *Science* in 1967. White argued that the attitudes that predispose us to environmental abuse can be found in the Judeo-Christian roots of Western thought. However, whether one agrees with his thesis or not, what was especially useful about White's essay was the realization it caused that the "environmental crisis" is, in fact, a cultural problem rather than merely a technical one. In short, it served to remind us that we *do* have tacit assumptions about the way the world works and how we should behave toward it, and in so doing this essay paved the way for the continuing examination which we now call environmental thought.

One can scarcely consider Renaissance thought without attending to what became known as the scientific revolution. I have dealt only marginally with this highly significant event, partly because it is so widely discussed elsewhere and partly because I tend to regard it as a dramatic elaboration of the basic transformation in thought which became apparent in the Renaissance. Some of the works mentioned above, such as Randall's *The Making of the Modern Mind,* deal with this aspect, but for a very readable introduction Hugh Kearney's *Science and Change: 1500–1700* (New York: McGraw-Hill, 1971) might be a good choice, as would E. A. Burtt's classic, *The Metaphysical Foundations of Modern Science* (Atlantic Highlands: Humanities Press, 1980). A longer history, equally readable, is Arthur Koestler's *The Sleepwalkers* (Harmondsworth: Penguin, 1964). Koestler deals principally with discoveries in astronomy, but his discussion relates to the overall understanding of the universe. A standard work with a more scholarly approach is Charles Gillispie's *The Edge of Objectivity* (Princeton: Princeton University Press, 1960). But perhaps the most significant discussion of the origin and effects of scientific thought is Alfred North Whitehead's *Science and the Modern World* (New York: Free Press, 1925), which, despite its date of publication, is still highly relevant. Steven Shapin and Simon Schaffer's *Leviathan and the Air-Pump: Hobbes, Boyle, and the Experimental Life* (Princeton: Princeton University Press, 1985) is a fascinating study in the struggle for dominance amongst ways of knowing nature, and Hans Jonas's *The Phenomenon of Life* (Chicago: University of Chicago Press, 1982) is a unique collection of essays, all of which provide invaluable insights into our understanding of living nature.

Finally, I have used a number of examples from the realm of art history, and the reader who wishes to pursue this topic should probably begin with E. H. Gombrich's classic, *Art and Illusion,* 2d ed. (Washington: Bollingen Foundation, 1961), and perhaps with his collection of essays, *Norm and Form,* 2d ed. (London: Phaidon, 1971), which contains interesting reflections on Renaissance art. Following that, the critique by Norman Bryson in *Vision and Painting: The Logic of the Gaze* (New Haven: Yale University Press, 1983) may be of interest, as might Samuel Y. Edgerton, Jr., *The Renaissance Rediscovery of Linear Perspective* (New York: Basic Books, 1975), William Ivins, Jr., *The Rationalization of Sight* (New York: Da Capo, 1973), or Svetlana Alpers, *The Art of Describing* (Chicago: University of Chicago Press, 1983). I have not dealt specifically with aesthetics, but any reader who is interested in that aspect of the experience of nature might consult Eugene Hargrove's *Foundations of Environmental Ethics* (Englewood Cliffs: Prentice-Hall, 1989), which also contains an excellent overview of the development of environmental thought.

The Liberation of Nature

The exploration of alternate understandings of nature is a challenging pursuit, and those who undertake it do so as concerned individuals rather than as practitioners of any particular discipline. Some do so almost as an afterthought, in the course of work principally concerned with other issues. For instance, Maurice Merleau-Ponty (*Phenomenology of Perception* [London: Routledge & Kegan Paul, 1962]) and Gaston Bachelard (*The Poetics of Space* [Boston: Beacon Press, 1969]) are both phenomenological writers who, in their descriptions, provide interesting examinations of the concrete experience of the otherness of nature. Edith Cobb (*The Ecology of Imagination in Childhood* [New York: Columbia University Press, 1977]) and Richard N. Coe (*When the Grass Was Taller: Autobiography and the Experience of Childhood* [New Haven: Yale University Press, 1984]) are both studying childhood experience, but in the course of that offer valuable insights into the child-nature encounter. And some, such as Max Oelschlaeger (*The Idea of Wilderness* [New Haven: Yale University Press, 1991]), include stimulating reflections along with an in-depth study of the existing ideas of nature. However, there are a few whose principal goal seems to be the illumination of the wild other behind the concept of nature, whether or not such an intention is explicitly stated.

Erazim Kohak's *The Embers and the Stars* (Chicago: University of Chicago Press, 1984) is one of the most personal and insightful accounts available, and one rich in philosophic insight. However, it is perhaps Kohak's literary or poetic ability that makes his account so vibrant, and in this he is similar to the other major contributors, among whom I would include Barry Lopez (*Of Wolves and Men* [New York: Scribner's, 1978], and *Crossing Open Ground* [New York: Vintage Books, 1988]), Charles Bergman (*Wild Echoes* [Anchorage: Alaska Northwest Books, 1990]), Gary Snyder (*Turtle Island* [New York: New Directions, 1974]), and Annie Dillard (*Pilgrim at Tinker Creek* [New York: Bantam, 1974], and *Teaching a Stone to Talk* [New York: Harper & Row, 1982]), as well as the earlier figures in nature writing such as Henry David Thoreau and John Muir.

Although writers such as these seem best equipped to describe their encounters, I must say that some glimpse of wild otherness is almost always present in the work of authors who are moved to speak for the natural world, whatever their occupation or academic discipline. To this, readers of John Livingston (*The Fallacy of Wildlife Conservation* [Toronto: McClelland & Stewart, 1981]), David Ehrenfeld (*The Arrogance of Humanism* [New York: Oxford University Press, 1978]), Paul Shepard (*Man in the Landscape* [New York: Knopf, 1967]), Joseph Meeker (*Minding the Earth* [Alameda: Latham Foundation, 1988]), or Richard Nelson (*The Island Within* [San Francisco: North Point Books, 1989]) can doubtless attest.

Index

About the Author

Neil Evernden was born in Vancouver, British Columbia, in 1943 and was educated in zoology at the University of Alberta, where he did doctoral research in "ecological aesthetics." He is the author of *The Natural Alien: Humankind and Environment* (1985) and has contributed articles to the *North American Review, Geographical Review*, and *Landscape* magazine. He lives in Newmarket, Ontario, and is an associate professor in the Faculty of Environmental Studies at York University.

WITHDRAWN